THE INTERVIEWING HANDBOOK FOR MILITARY LEADERS

PUBLISHED BY:
KCE PUBLISHING
P.O. BOX 1539
SAN ANSELMO, CA 94960

printed in the United States of America

TABLE OF CONTENTS

TABLE OF CONTENTS

TABLE OF CONTENTS

PREFACE

This comprehensive Handbook has been complied by an experienced and qualified resource in order to prepare conscientious military leaders who anticipate interviewing for civilian career positions, and whose ability to present their qualifications and credentials in a professional interaction will directly relate to the success or failure of the aforementioned efforts.

Which is to say that this Handbook has been written for just about everyone who is in the military now or has been a part of it during the recent past, and needs to understand the rules by which the "outside" employment world operates in order to successfully find a job.

The difference between the two paragraphs above typifies the transition you may need to undergo in the near future. The meaning of the two statements is nearly identical, but the style in which they are delivered is wholly different. You are the same person today as you will be when you face the civilian world in the near future. Yes, your hair might grow a little longer and your clothes may change in some ways, but inside, you are the same. You have been operating in a structured environment which, if you are lucky, has not left you with a stereotypically militaristic clipped, formal delivery and a cold, uncompromising personality (see paragraph one). If it has, your efforts to become less formal, less clipped (see paragraph two above) and more empathetic to the situations and people around you must be double that of the person who has survived the military experience with a clear perspective of self and who understands how minor a role a military leader plays in the larger "world" of which you are about to become a part.

This new edition of *The Interviewing Handbook* incorporates some important changes from earlier editions. The Handbook was first published in 1976, and has gone through many changes in the years since. The process of recruitment and selection in our nation's workplace reflects fundamental changes in our society in general. As a result, the interviewing game changes, subtly in some ways and drastically in others, to conform to new trends. This edition represents the most current processes and practices in the employment field, and when those change, a new edition will be published to reflect the facts.

We are going to have some fun while preparing you for the interaction known as an "interview." I have written this Handbook in a casual and easy style. The employment interview is a situation that calls for humor, relaxation and natural personality flow (is that news to you?), and your preparation for interviewing should be the same. Our purpose is to prepare you thoroughly, so that when you are actually in one, you will be successful beyond your greatest expectation.

Sometimes I get carried away by my enthusiasm, and certain sections of this Handbook contain statements which, if taken out of context, may read as if they were laws or rules, never to be broken. That's because I feel strongly about them from my own experience as a recruiter, an employer and a consultant. **Please understand: there are no rules that are absolute. Beware of anyone who tells you that A equals A in all interviewing/employment situations.** There are no absolute rules for the simple reason that interviewers are humans (although you may find yourself wanting to dispute this fact later), each with their own viewpoint and bias, making definite or absolute predictions about their behavior impossible.

You will need to forgive my emphasis on business (vs. non-business) interviews. All the references to corporate situations in this Handbook apply equally well to institutional or non-business interviewing situations and will bring similar successes.

You will get the most benefit from this Handbook if you read it with a pencil in hand. Make notes, do the exercises, mark it up. It is a tool to be used, not just to be read.

FIRST,
A SHORT EXERCISE

The major reason you purchased this Handbook was to learn the techniques that would allow you to interview successfully for a career position. I have created a yardstick so that you may measure your progress and be able to determine whether your investment has been a good one. Here we are in the very first section, already talking business!! What's an acceptable return on investment for you?

Below you will find a list of fairly standard questions. They are questions you can expect to hear in most employment interviews. Perhaps you have already given them or some like them a little thought. I would like you to picture yourself now in an interview for your ideal job, one you really want.

Answer these questions. You don't need to write out answers completely, but you should make some notes on the following pages —an outline of your answer—, so you can refer back to it at various stages as you use the Handbook. Remember: you are in an interview, trying your best to get an offer for that job you really want.

1.   Where did you grow up?

2.   What are some of your favorite hobbies or pastimes?

3.   What attracted you to this particular opportunity?

4.   What are your salary requirements?

5.   What geographic location do you want to work in?

6.   What do you consider to be your most outstanding personal attribute?

7.   Tell me about any sales experience you may have had.

8.   If your most recent commanding officer could have changed anything about you, any personal characteristic, what would it have been?

9.   What did you gain most from your military experience?

Have you written outline answers to the questions? I think that nine questions should be enough for the purpose of evaluating how you might change your answers, if at all, once you have been exposed to the information outlined in *The Interviewing Handbook for Military Leaders.*

Many articles in newspapers and magazines address the subject of interviewing superficially, telling you to analyze who you are and what you want, and few give more than general guidelines on how to accomplish this important step. The unique aspect of the Handbook you are holding is that the entire book can be used as a self-analytical tool, if you take the time to answer the questions, do the exercises, and apply the insight you gain to better understand yourself. It is a somewhat uncomfortable task, but essential. Without a complete understanding of who and what you are, you will have difficulty marketing yourself. Period.

Even though the entire Handbook can be used to help you know yourself better, this section will outline some specific exercises to help you get your bearings. Be brutally honest with yourself as you do these exercises. I would strongly suggest that you isolate yourself, and be prepared to spend an uninterrupted hour or two. Most likely these exercises are personal in nature and you really won't want to share the results with others. There is something to be said, however, for an outside opinion. If you know someone who you feel comfortable with, and who knows you well, you might ask him/her about some of these exercises. You may find that there is a different view from the outside looking in. All that said, let's go. I've supplied the paper, you bring the pen and coffee.

The first exercise is the triangle evaluation, a method used frequently (but not always called the triangle) to objectively evaluate a candidate who has attended college in the fairly recent past. This evaluation is included in the Handbook because many of its users are junior military leaders who have graduated from college, then served in the military for 4-6 years — prime candidates for the triangle evaluation.

There is a factor at each of the three corners of the triangle.

- On the first, the grade point average (GPA) you earned in college.

- On the second, the number of hours you worked each week at a part-time job during the school year.

- On the last, the approximate hours per week spent involved in extra-curricular activities.

The last two corners involve some estimating and averaging because you are calculating over a period of four years and probably had an inconsistent level of employment and activity. But give it your best shot. Find those numbers honestly, perhaps with the use of your calculator. This is an objective exercise and there are no corners on that triangle for school status, toughness of curriculum, pleasing personality, or any of that kind of evaluation. This is not meant to be a flattering exercise, but rather an honest evaluation of how you applied yourself to the goal of graduating from college. Fill in the numbers.

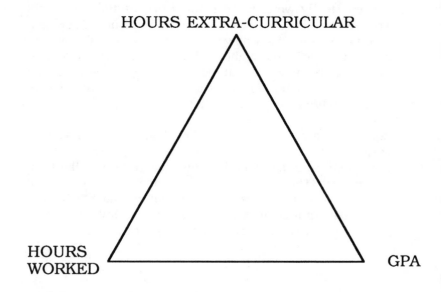

HOURS EXTRA-CURRICULAR

HOURS
WORKED

GPA

If you have high numbers in all the corners, you are showing very positively in this evaluation.

If you have a low GPA, but rank high in the other two areas, you are still doing OK, but there may be some questions that need to be answered.

If you have a high GPA, and one of the other two points has low numbers, you are still on semi-solid ground, depending upon your answers to some questions.

If you have a high GPA and both the other points are low, some major questions are raised about what progress you have made in your capability to relate to people, show leadership and about your total education (in contrast to your book education).

If you have low GPA numbers, low hours worked, but extremely high extra-curricular involvement, questions are raised about what progress you have made in setting priorities and determining purpose.

If there are high numbers of hours worked, low GPA and low extra-involvement, questions arise concerning the hows and whys of financing your education. Perhaps some exploration of other areas such as the breadth and depth of your educational experience is in order.

Or, if you have low numbers in all corners, serious questions arise concerning what sort of progress you have made since then in the areas of maturity, application of self to purpose, intelligence, and personal interaction. Someone who has not worked or participated in extra-curricular activity and graduated with a 2.1 GPA will have some explaining to do. You will note that I have not said that any of the classifications above are "good" or "bad". No recruiter worth his salt would make a decision based solely upon the triangle evaluation, especially when evaluating candidates who graduated several years ago. There is much more to take into consideration (not the least of which is your military training and leadership), but this is one way to arrive at a fairly objective evaluation of how the candidate has focused energy, and in what priority, on the task of graduating from college.

What if you are 42 years old and can't remember what college you went to, let alone what your GPA was? Does this triangle apply? You bet it does, but the corners become different factors. How's about salary, complexity of management role, and community involvement or avocational interests for the three factors? It's doubtful that you will ever escape the triangle, and it becomes even more important as you advance up the ladder. Remember the triangle and the balancing of priorities it evaluates when you are interviewing for the position of chief executive officer of a corporation.

It may seem to you that there is some subjectivity in the triangle evaluation. Well, there is, but it is all even more subjective from here on in the self-analysis and interviewing process.

Below and on the next page, list ten strong personal qualities or characteristics that you feel you possess. Pick positive ones, and be perfectly honest with yourself. If you only list qualities you think sound good, or ones which you think a recruiter would be interested in, you are not going to get much out of the exercise. Take the time to list all ten before you read further.

Now that you have listed ten outstanding personal qualities, next to each of them write an example of when the quality helped you to achieve a certain goal in your life. Example: you may have listed a personal quality as *competitive* Now cite an example of a particular time when that attribute helped you to achieve a goal, such as: won the first chair in the violin section of the college orchestra. Go back to the list of ten and do that for each of the personal attributes you have listed.

The reason I had you list the qualities first and then the goals those qualities helped you achieve, was that I want you to test whether the qualities you listed are important ones, as related to your achievements in life. If you came up with *honesty* as a personal quality , but cannot think of when it really helped you achieve anything, this might not be a good quality to pick. This does not mean that honesty is an undesirable trait, but only that you are going to have difficulty explaining why it is important relative to your achievements. Choose a different one.

Let's create another list on the next page. This time list ten things you do exceptionally well. Try not to be pornographic, but come up with ten abilities you can honestly say you do better than most of the people you know. The list of abilities may or may not be related to the ten qualities you wrote earlier. Example: I can shoot a back azmuth better than anyone I know. Go ahead and make your list.

When you finish the abilities list, there is one more list to make. You are going to list all the major accomplishments you have achieved during the time you have been making independent decisions in life. Some examples: buying a new Porsche, being elected as class president, getting married, earning a commission. If you do this well, you will find that the list is longer than you would have expected it to be. Do a mini-review of your life. When you finish the list, on the next page, continue reading.

Now write a few words describing the **real** reason you achieved the goal, next to each achievement.

Some examples:

became class president .... peer pressure

received commission .... father

bought a Porsche .... status, investment

got married .... security, social pressure

Interesting, isn't it? Maybe the truth hurts a little, but be honest with yourself. It is far better to discover these facts here than in an important interview.

Now let's go back to the first list you made, the list of positive qualities. Look them over. Add to that list at least four things you think you need to improve. Examples: shy, have temper, need to lose 20 pounds. You are never going to talk about these things in an interview, but you should be aware of them, and face them in black and white.

Now, looking at all the lists, can you draw some conclusions from them? For example, is there a repetitive attribute you see in them all? Perhaps a consistent need, desire or ability to relate to people? To avoid people? Or perhaps the orientation is to attention to detail, to the need to be out-of-doors, to be competitive, or to be non-competitive, for that matter. I leave the bulk of that analysis up to you. You know whether or to what extent you have been honest in making these lists. If you can draw some conclusions, find some common denominator, it might point you in the direction of a career that uses those strong qualities or fulfills those strong needs. I feel that someone needs to know you really well to give good career advice. Since you know you better than I do, you will need to do most of the work.

If you have absolutely no idea what sort of career you should be focusing upon, it is time to get some professional advice. While we generally refer to sales/marketing, finance/administration and production/engineering type positions in this Handbook, obviously thousands of variations exist out there. A knowledgeable career counselor will cost some money, but could be easily worth the expense if your path becomes a more certain one.

One thing you will be encouraged to do as you work through this Handbook is to address the **purpose** of your career. Why are you going to work? What do you expect to get out of it? How much are you willing to give up to get it? Companies pay salaries, some larger than others. The company extracts energy and time from you in proportion to the salary you receive from it. After all that energy is drained, is there enough left to enjoy life? The intent of the exercises above was to get you thinking about who you are. They should have taken some effort and about two hours' time. If you did them, you are already several steps ahead of the competition. If you skipped over them, you are cheating yourself. Go back, spend the time and do them.

One of the first decisions you will face in your choice of employment is whether you should focus on working for a large organization or a small one. A discussion of the pros and cons of both is also excellent interviewing material. It gives the interviewer certain information about you: your values, your level of understanding, and your needs. Before we go into detail, let's define our terms.

A major national company would be any company with offices and representatives throughout the country; it has activity in all or most of the United States. Towards the other end of the spectrum would be a regional company, smaller and restricted to a specific area. Although this does not mean absolutely that the regional company is smaller than the national one in terms of income dollars or employees, these factors are generally related to scope. The classification I seek to differentiate here is the boundaries of geographical operation **and** the related size of the operation.

What are some of the advantages and disadvantages of working for a large, versus a small, organization? The following is flavored with generalizations, understandably:

ADVANTAGES LARGE

<u>Extensive formal or informal training programs.</u> Major organizations have the time and money to train you in your job, even to the point that you may feel over-trained. As you will see, there is an image at stake here, and no tolerance for an under-informed representative in the major company.

<u>More management opportunities.</u> Within the framework of a large organization, you can often move around and pursue opportunities in different departments or divisions. Being blocked from promotion by a young manager is less of a problem when there are other opportunities to pursue.

<u>A system, with rules, within which to operate.</u> An established company has become so through consistency of operation over

21

time. The corporate policies and systems are in place to reflect that consistency and, while hardly inviolate, represent the boundaries within which you will be operating. These rules are different for all companies, but have some commonality among the large national organizations.

Extensive benefits. A large company generally has more expensive and wider-ranging benefits. Although in recent years the cost of some of these has drawn notice from budget slashers, you can still expect fairly generous medical, dental, retirement and other benefits. The more significant benefits are stock purchase plans, savings plans that are matched by the company, stock giveaway plans, free insurances, low cost insurances, profit sharing and other goodies. Small companies simply cannot afford them, or choose not to spend their money on those benefits.

Bigger paycheck. In the past, I was convinced that smaller companies offered higher starting salaries than large major companies, and that may often still be true. Smaller companies may have to "buy" topnotch talent by offering more on the front end in some cases. However, as a general rule, I believe that large corporations compensate better over time than smaller corporations, particularly when fringe benefits are taken into consideration.

Well-known name. This might seem trite to you, but after the nine millionth time you have explained that Acme Manufacturing makes ball bearings for baby buggies, you may see some of the value in a Fortune 500 name.

ADVANTAGES SMALL

Smallness of size increases chances for recognition at top. If the company is small enough, everyone knows everyone, and often by first name. If you are doing well, you will get noticed by those who make the promotion decisions. This is both an advantage and a disadvantage, obviously, because if you screw up, everyone notices that, too. The power to influence promotion decisions generally rests in fewer hands in smaller companies.

Incredible growth. Small companies get large quickly, if conditions are right, and you can ride that wave if you are in the right place at the right time. There is a risk-reward balance here, as anyone in a computer-related industry can tell you. The success stories are breathtaking; so, too, are the "crash and burn" stories.

Less relocation, or only within a certain area. Read the section below under "disadvantages large" for more information, but generally small companies do not relocate their personnel as a matter of course with the same frequency, or across the geographical span that large ones do.

Distinct sense of purpose and dedicated effort. Whoa! I'm on dangerous ground here! This is not to suggest that employees of large corporations lack a sense of purpose, but rather that small organizations cannot compete without it. The majors are discovering that the energy created when all members are functioning as a dedicated team can literally move mountains, and they are applying it to their businesses (more about this later in the section titled "Your Greatest Challenge"). What is interesting is that a variation of the same theme has been playing in smaller entrepreneurial businesses for a long time with little notice, and is essential to their very survival. Being a part of that can be a distinct advantage.

Quantum leaps in responsibility. There are generally fewer managers per capita in small companies, each with broad responsibilities. Jumping one level in a small company may bring you to a scope of responsibility only achievable in three or four jumps in a major organization.

DISADVANTAGES LARGE

Impersonal treatment. The enormity of some major organizations can make the individual employee feel like a number in a computer. It is not quite like working for the Federal Government, but the impersonal is reflected in employee code numbers, computer-generated "congratulations," inflexible rules and ID badges. In fairness, many companies have recognized this detachment as harmful and counter-productive and have introduced processes for change. But those processes are only as effective as the people who use them.

The systems that are an advantage (see above) can limit flexibility to do what you want to do. Living within the system means that you cannot move outside of it without the consent of the powers that be. True, you can get consent by showing how the exception will bring a return on investment, but the flexibility to do so more or less unilaterally does not exist in large, established corporations.

Almost always relocation when promotion. The chances are great that a large company will promote you and move you to another location. There are exceptions. The costs, financial and otherwise, of moving employees are getting higher daily, and companies are sensitive to costs. Some will try to relocate only regionally, or not at all when possible, but you are wasting your time if you go to work for a national corporation with the intent of staying in one place. Why? Because the opportunities are fewer as you go up the ladder. Consider this scenario: if rung #1 is in Toledo, there are probably sixteen #1's in Toledo. But rung #2, that's different. There may be only two of those in Toledo, but there is an opening in Memphis for a rung #2, so you are on your way. What!? You don't want to go to Memphis!? OK. You just sit there on your rung #1 in Toledo, and we'll call you. Maybe if a #2 comes open in Toledo, and nobody in the entire system either wants it or is ready to go into it, we'll offer it to you. Maybe. Rung #3 is probably in Chicago, and #4 in Dallas, to say nothing of the parallel moves you may have to make to gain knowledge in a different market or region. Think about it. If you want to live in Toledo the remainder of your life, you'd probably be better off with a Toledo company. Yes, it is possible to work for a major company in only one location, but you will be passed up for promotion, be relegated to non-developmental kinds of jobs and given less overall responsibility than the person who will go where needed. There is a high price to pay either way and any one decision may be right for one person and not the other. You owe it to yourself and your family to make the decision consciously and now, not after six years of employment.

DISADVANTAGES SMALL

Training programs limited. This is the flip side of the expensive training that the majors do. I've seen sales reps from local companies sent out on the road with a map, a catalogue

24

and an order book. The resources are often simply not there to spend on the individuals. For that reason, many smaller companies prefer to hire personnel with some experience in their industry.

Limited fringe benefits. Benefits are incredibly expensive. And unless the small company has publicly-traded stock, some stock-related benefits may not be available at all.

Relatively unknown. This can be particularly frustrating when making career moves. The instant recognition for quality of training, etc., that many of the nationals enjoy will be lacking when you go out with your resume reflecting Acme Manufacturing. You will need to sell harder.

There, I think that has been said as objectively as possible. If you detected a slant toward the big national companies, it's because it exists. For the person going into a first career job, it seems to make sense to get good training and work for a company with a known reputation and then, if one does not want to tolerate the negatives associated with the majors, go to work for a small company where there may be different negatives. The small company should welcome you with open arms since you know all those big-league customers and have had all that high-powered training. The reverse is generally not true; most majors will have little interest hiring someone who has worked for our old friend Acme Manufacturing down the street. They will need to re-train that person to do the job "right," and that may be more work than it is worth.

I may have just blown the idea you have that I have heard countless times from people entering the job scene: "I'm going to work for a small company for a couple of years, get the experience and then go to work for a large company." I'm not saying it never happens, but it happens with less frequency than those newcomers expect.

NOW LET'S TALK ABOUT

SECURITY AND STABILITY

Paychecks from large organizations tend not to bounce as frequently as those from small ones (although I can think of a

25

couple of noteworthy exceptions). That's about the only thing you can relate security to in this day and age.

We live in an exciting stage of business evolution. The headlines are filled with news of contracting and expanding businesses, sharks and knights, leveraged buyouts and mergers, bankruptcies and Chapter 11's, parachutes and takeovers. Many of the terms didn't even exist when the first edition of this Handbook was written in 1976. Things are changing fast, and the sense of stability and security our parents understood to be part of the reason one worked for a big company has dissolved, or at least been diluted. We all value some sense of security, but if you are someone who really needs it badly, someone who is terrorized by being out of a job, do yourself a favor and stay in the relatively secure military. Now, more than ever before, stay in the military. The **only** security that exists out here is your own firm personal conviction that you are the best you can be at what you do, and that your skills are marketable. You will not find security in a company. Some industries are more stable than others, larger companies are sometimes less vulnerable than small ones, but all are subject to the changing forces that are in play currently. It's a real jungle out there, as someone has already said, and if you fear the tigers, you'll be eaten up.

Large company, small company. You will decide which of these two is the most attractive for your personality and situation. As you will see, you can interview for both situations at once, but you must understand the differences to make a good presentation.

The sociological forces that have changed the business world, as well as our society, over the past fifteen or so years have almost made the points around this subject moot. What has happened, of course, is that the hiring managers who had the mindset that anybody over twenty-six years old and single was either a shifty no-account who would skip town with the petty cash at the first opportunity, or, worse, a homosexual deviate who would bring the company to ruin and the headlines of the *Wall Street Journal,* moved off to the great retirement community in the sky to be replaced by younger people who had different perspectives. This younger set generally knew that married people can also abscond with the petty cash and many men and women are, literally, not what they appear to be.

However, I still think there are valid interviewing points of discussion around this subject of being married or not married, and at the very least the old values need to be discussed in the off chance that you meet head-to-head with someone who still holds them. As you go up the interviewing ladder, the age of those at the uppermost rungs is up there too, and people have a very human trait of keeping their old prejudices in cold storage, just waiting for the opportunity to drag them out and use them.

First, let's establish that your marital status is **not** a subject that a recruiter can **legally** ask about. The question "Are you married?" violates Federal discrimination laws, and while we generally think of these laws as set up to protect minorities, they apply to all people in our society, including the white male. While the question is illegal, it is only the thickest of recruiters who will not pick up marital status information in a normal interview. When you refer to "my wife" or "my husband" you make the point obvious, as well as when you (if single) say, "Since I'm not married, I can relocate tomorrow." It is illegal for a recruiter to ask it, but it's not illegal for you to offer the information, and there are situations where you will want to offer it.

Your marital status is related to an area that is of valid concern to an employer: your flexibility. I'm going to devote a lot of pages in this Handbook to relocation, since that subject is a major factor in winning the corporate game and getting promoted to higher levels of responsibility. Simply put, if you are single and living in an apartment building, it is going to be far less costly and probably a lot easier to get you to move to Phoenix than the person who is married, living in a house and has two children in school. It is a valid concern, and one that becomes illegal only when the recruiter decides **for you** that you wouldn't be interested in that relocation. So don't allow the situation to get to that point. Tell the interviewer that you can, you will, and you expect to relocate in your corporate career. As you have read, I feel strongly about the rest of this: if you are not willing to, don't have the ability, desire or expectation of relocating, think long and hard about interviewing with major national corporations. Period.

Another couple of thoughts about your marital status and employment opportunities: if you are applying for a position that requires a heavy travel schedule, it is possible that you would think a company would show a preference for a single person. Depending upon who is doing the hiring, that may be true, and if you are single, you have an asset you'd better get on the table and use. Mind you, no one is going to tell you that the preference even exists; you need to determine that from the facts at hand. In the same manner, if you are in competition for a position which requires a high level of social contact, with entertaining in the home or at other social events as a common occurrence, you'd be wise to let the company know that you have a spouse who is well-suited to help you carry out that obligation.

In the not-so-distant past, people tended to believe that married people were more stable than unmarried people, and were preferable as employees. Given the current divorce rate, I doubt anyone could make that statement with firm conviction, but I bet a lot of the older generation still have that idea in cold storage. In spite of the fact that the person who is married and "stable" when hired could easily be single six months later, I guess you could make a case that the person who is married with three children, a $120,000 mortgage, two cars, a boat and a puppy might be less inclined to tell the boss to take a flying leap than the single person whose obligations stop at the monthly rent payment. But I wouldn't want to bet the farm on it.

As long as we are deep into this topic of marriage, let's take it one more step to the point where you are married to a person who also has a career he or she would like to pursue in a corporation or other organization. The dual career issue is a relatively new one, and most corporations are still sorting out how they will handle it. Unfortunately, these sorts of things take much too long to resolve in our society. Even with the enormous number of two career families at work today, business America has yet to deal with some of the fundamental issues that come out of that fact, such as on-site child day care and flexible workday hours. Likewise, the issue of promotion and relocation for dual career families is seldom addressed directly and officially.

If you are an accountant by training and your spouse is a mechanical engineer, there is little doubt that both of you could be employed in most major metropolitan areas, even within the came corporation. Some companies have policies which would prevent both of you from being hired in the same unit or organization, and other companies don't care much, except where reporting relationships cause issues. At any rate, the accountant/engineer couple could find itself gainfully employed in the same or different companies. Most of the couples I have met are not engineers and accountants, but engineer - secretary, sales representative - nurse, airline pilot - designer, marketing manager - buyer. In most of these cases, relocation for one partner can cause major problems for the career of the other.

Since the issues here involve dual careers, and dual careers produce dual incomes, the main focus may be financial. The unplanned loss of either partner's income through relocation could cause major financial problems. The key word here is "unplanned." When the interviewer wants to know how you and your spouse are going to deal with reality when you get hired or promoted and sent to Urbana, Illinois, all that is expected of you is to demonstrate that you have discussed the issues with your spouse and that some sort of alignment has been reached and that the two of you represent a unified entity, capable of making the right decision once an offer has been received.

Actually, this dual career/relocation problem is another reason former military leaders have an advantage in the hiring process. Most likely you have moved around in the military enough to have surfaced these problems before, and have reached some sort of family agreement. Perhaps this has meant a mobile career for one of you. Tell the recruiter and use this experience to your interviewing advantage.

Most importantly, if you have not done so already, you owe it to yourself and your family to get this dual career/relocation problem on the table for discussion **now**, before you get involved in interviewing. If you are **assuming** that your spouse will carry on in the same manner he or she has in the past military years, you may be in for a rude awakening. If you are thinking about making a compromise with your spouse to accommodate both your careers with one living in Dallas and the other in Memphis while commuting to one place or the other for weekends, I strongly suggest you find someone who is doing or has done this, and spend some significant time investigating the drawbacks. I'm not saying it cannot be done successfully, but those have got to be some very difficult times. If you do decide to do it, **never** admit to that fact in an interview. First, it is no company's right to ask about it. Second, it may eliminate you from competition, based upon the potentially high failure rate of such arrangements and the personal costs associated with those failures.

This whole married/single subject needs a final point of discussion before we move on. Many people who are recently divorced or separated, an ever-increasing percentage of our population, seem to have an incredible need to talk about their separation or divorce. If they are interviewing at the time, they talk to the recruiter about it, in detail. You wouldn't believe the detail. It is understandable; a good recruiter listens well and gives you an atmosphere in which you feel comfortable talking. But please, do yourself a favor, talk to a professional counselor, psychiatrist or friend about your marital problems, not to the recruiter. Furthermore, since the subject is off-limits, legally, think of yourself as one of two classifications: you are either married, which includes separated, or you are single, which includes being engaged, and if you have never been married before or your divorce is final. You are not "kinda" married in a job interview. All recruiters are wary of candidates who are about to go into or are in the middle of a divorce, as even the most amicable are emotionally staggering. Don't offer this potentially damaging information to an interviewer on your resume or in conversation.

I know it is illegal to discriminate against you because of your marital status. It is also illegal to smoke grass or drive over the speed limit. Most recruiters can interview candidates and truly disregard this information, but there are many who cannot. All our values are the result of our experiences in life, and obviously those vary from person to person. Since you will know little about the person with whom you are interviewing and cannot discern that person's values on the subject, be conservative with information about your marital status.

32

Originally, this chapter was entitled "The Junior Military Officer." After several editions, and a detailed marketing analysis of who was actually buying the Handbook, I decided to change the title of this chapter, as well as the title of the entire Handbook in order to encompass the wider audience it has enjoyed since 1976.

The main thrust of this Handbook is toward the individual, male or female, who has entered the military, invested two to six years learning to and becoming a leader and, at the point this Handbook is being read, is preparing to leave the military and begin a civilian career. This military leader has developed some unique personal attributes that are very desirable, especially when contrasted to the non-military person in the perspective of hiring into what are mostly entry-level positions within the civilian employment environment. If you are not in the military, have never been, and don't expect to be (although that is never an absolute if you are healthy), read this section anyway, for the military leader is your competition at the level you may be trying to enter the business world.

Some of the qualities that make you stand out as a military leader are probably obvious to you: you are somewhat older than your civilian competition and, usually, have some maturity to go along with that age. You have enjoyed, or endured, responsibilities far greater than those of your civilian age group (did you ever go home for leave during your military career, meet an acquaintance or classmate who had not gone into the military and just wondered at the immaturity you saw?). Within the framework of the military, you have been a manager. All along, as you have put up with all the incredible restraints and bureaucracy, you have known that these skills you have been developing would be valuable assets someday. And now you are looking forward to the approaching situation when you can apply them to a civilian position. Fine. But there are other, perhaps less obvious, assets that also make the package attractive to the potential employer. Let's look at some of them:

-You have probably developed some people skills. That is, you have had the opportunity to tell someone, or a group of people, what, when, where, how and why they should do a certain task; (and if you are lucky) you have learned to do that in a way that makes those people want to do it. That experience is worth plenty, because you have made your mistakes in the military, and have learned from them.

-In most cases, your relocation costs will be paid for by the government. There is more detail about this subject in the section "Who Is Going To Move My Stereo," but basically it means that you can save the company that hires you many dollars if, instead of your civilian counterpart, you go to work for them. Yes, yes, most major companies pay relocation costs for new hires nowadays, but your military move is a definite asset. Make sure you know what you are doing before you use it.

-If you are married, and often military leaders are, you and your spouse are familiar with packing up and moving out with a few weeks' notice. Your civilian counterpart may never have had to do that before. This is not an insignificant asset given the importance of relocation in the corporate world.

-You have learned to live within a structured environment. When the boss says, "Let's do it," you respond with vigor, determination and a sense of purpose. Never underestimate the value of this lesson you have learned. Corporate environments are more participatory than they have ever been in the history of industry, but when you get right down to it, after the discussion has been held and the decision has been made, those that can and will do are the ones that are valued the most highly.

-Finally, you have done all this at a relatively young age.

If you put all those things together, you may be looking at a very marketable product. I use the word "may" because I don't want to give you the idea that just because you are a military leader, you have a free ticket to the stars. That is hardly true. In fact, the best you can expect is that some of the organizations you interview with will give you a slight preference in the situation* (or maybe you will be working uphill if you meet an anti-military interviewer). But you must still be competitive with your civilian counterpart. In addition to all of that - and this may be hardest for you to accept - when you do land your offer, you may find that the salary and position are identical to or just slightly higher than that of the college graduate right off the campus. So here you are, two to six years down the road from college, having served our country, and I'm telling you that you will probably receive the same salary and responsibility as the person who hasn't done any of that! Just where is the justice in that?! Well, I'm going to tell you, but first settle down and develop the right attitude.

Yes, you have done more than that 22-year-old right out of school. We detailed all that above. But you must admit that your experience, while related in part, is largely in another field, another world, than that of any business or other civilian position for which you are applying. Some employment opportunities such as Civil Service and other government jobs give you direct credit for your time in the service. That makes sense, you are operating within the same bureaucracy. If that's right for you, that's great! In other worlds, you are still an untested employee, unless you are making a parallel move to the civilian sector from your military job. Accept that.

Financially, you had better be prepared to take a cut. The whole of civilian employment has done a fairly good job of keeping up with military levels of pay, but you may be paid more as a military leader than the economy can bear. Cutting back is never easy. If you are still a few years away from leaving the military, you'd be smart to start saving now to make up the difference when you do get out. Expect a cut. You may find that the forces that drive the salaries in the open civilian job market exceed those forces (read Congress) which drive the military pay increases, and you will be delighted to have all that money in the bank to spend on BMWs and other

---

* if you are working with an employment firm specializing in the placement of former military personnel, you can expect greater enthusiasm for your background

trappings. That would make your transition easier, wouldn't it? Remember, your fringe benefits will generally be exceptionally good and better than military ones, in spite of all the publicity the military puts out to the contrary.

Enough already with the unhappy news. Let's detail how you can expect to benefit from the investment you have made in the military. Aside from sometimes being given a place in the front of the line when you are interviewing with an organization, more is going to be expected of you in everything you do. This means that you will do the same job as a 22-year-old college graduate, but you will be expected to do it better because of your maturity and experience. "This is a benefit?" I can hear you asking yourself.

Yes, it is. If you look at the promotion rate of recent college graduates compared to that of former military leaders who enter business at the same time, you will see that the military person gets promoted at a faster rate. You would also find that the military person gets more frequent raises and has the flexibility to be positioned in a wider variety of jobs. There are exceptions, of course, but the former military person has already made common growing mistakes in the military, has developed important capabilities, and usually does a significantly better job. I will tell you repeatedly in this Handbook that you, the military leader, are considered to be prime material for civilian employment opportunities. A major reason that is true is in this paragraph. Read it again if there is any doubt in your military mind about the reason why.

In contrast to your experience in the military, civilian employment successes and promotions will come because of your **ability** and your **productivity** in the work environment, **not** because of the length of service you have racked up for the company. Let me tell you of a real case in point: I have a friend who served in the military with me. Immediately upon separation he started working for a subsidiary of a major national corporation as a sales representative. Two years later, he was national sales manager for the subsidiary with approximately 60 sales and sales management personnel reporting to him. He went through four promotions in two years, and in doing so, he was promoted past employees with many years of service in the company. But he did a better job, and was promoted for his ability. Yes, he'd be the first to admit that he had some lucky breaks; he did the right thing for the right people at the right time, and was in the right place when the opportunities hit. But the fact remains that in two years, he went from level one to the top of the sales force, the equivalent of going from second lieutenant to commanding general of sales, in about the time it normally takes a military person to advance from 2nd to 1st lieutenant in uniform.

This transition is not going to be an easy one for you, but I'll bet that you will look back on it in a few years as being a highly exciting and rewarding time in your life. As the date approaches when Uncle Sugar cuts off the paycheck, your anxiety level will rise, although you may not admit it, preferring to focus on the excitement of "getting out." If you have had no other way to judge the job market than reading the overseas edition of the *Stars and Stripes* and an occasional hysterical letter from home, you can be assured that after reading this Handbook and doing the exercises, your interviewing skills will help you master and overcome whatever competition is awaiting you.

Want to start a lively discussion? Ask people whether they feel college degrees are worth the time, trouble and expense it takes to get one. Or, better yet, ask the same question, but about Masters Degrees. What you will discover, and quickly, is that the subject is an emotional one, and people will almost always have strong opinions. In spite of the reaction it may produce in you, I feel it is appropriate to confront the subject.

Most civilian organizations hire personnel on three or more levels, just as the military organization you have been a part of does. In the military, it is easy to differentiate among those "hired" (officer, warrant, and enlisted levels). You wear that badge on your uniform, for all to see. Very clear and above board. Civilian organizations also hire on at least three different levels, but the approach is much more subtle to the outsider than the insignia on the collar. Oh, the insignia are there, but you need to know the code, and that may take several months or even years to understand.

The point that I'd make is that your level, the place you are hired into in the organization, is often directly related to your educational background.

Let's start out with what we will call a Level One position. At this level, the employee is most likely a non-degreed individual. The position could be clerical in nature, or on the production line, or possibly even in the sales force. The point is that it does not have any definite planned progression from this level to Level Two.

People who fill Level One positions are generally expected to remain in that job for a long time. Their "unpromotability" can be the result of several factors working against them, the most obvious being the lack of a college education. Other more important factors may be the lack of visibility these kinds of positions afford, the lack of real impact and responsibility inherent in them, and the sheer numbers of these kinds of positions.

Understand, please, that this is not necessarily a bad situation for either the employee or the company. Not everyone can be a manager, or wants to be. Often these positions have security, status and good salaries. A military example: the non-commissioned officer who has seen lots of commissioned officers come and go, trained them, if they'd admit it, who knows the job well, and is the backbone of the unit. The civilian equivalent may be called salaried, non-exempt (meaning that personnel in these positions are not exempt from the labor laws requiring payment for overtime and other benefits), or are called hourly or wage personnel, sometimes contracted to perform a specific job at a specific rate of pay.

This lowest level, then, is non-developmental, only in the sense that there is no real established track to take one to a higher station. Does it ever happen anyway? You bet it does, all the time! But promotion to the next level occurs more because of luck and hard work than organizational design. If you are expecting to be a mover and a shaker, this level could slow you down incredibly.

The second level is the one most college-trained candidates shoot for: the developmental position. This position could have all the **outward** appearances of a Level One non-developmental situation, but has a planned career progression which takes the employee down the path to higher levels of management. The difference between a Level One and a Level Two position is the difference between a technician and an engineer, a bookkeeper and an accountant, or a sales clerk and a sales representative. **However**, it could also be the difference between an engineer and an engineer, an accountant and an accountant, or a sales rep and a sales rep. The confusing truth of the matter is that there are "engineers," "accountants" and "sales reps" who have those titles, who make those salaries, who might even have those degrees, but who are going **nowhere.** They are filling a position and doing the job adequately, but have failed to make the cut for higher levels of responsibility. Many people are satisfied with this station, and will retire from those jobs. It is not a bad situation for the company or employee, so long as the company has sufficient management potential personnel in some of those positions. People who are not satisfied with this station usually leave to seek a better opportunity elsewhere.

In short, Level Two is an opportunity position, one in which the employee is expected to develop and advance. However, if the employee fails to demonstrate the acceptable level of managerial talent, the position can become a Level One position, and the employee can advance only with extraordinary effort or success.

The highest level of hiring takes place in what has become known as "the fast track." At this level, the ballgame is played with different rules. Players can expect to experience faster advancement, higher pay, higher visibility, and the higher failure rate that goes along with all that. The rewards and the risks are great. You either go up or you go out.

At this level, the educational requirement is often an advanced degree. In technical situations this means a Master's or possibly a Doctorate in engineering, education, science, etc. In the business world, this translates into a Master's of Business Administration (MBA), and usually one from the top business schools.

Since it is likely that some readers of this Handbook would be attracted to this Level Three, regardless of the risks involved, let's discuss Master's Degrees a bit before you all rush out to get MBAs.

While it might be only rarely that a non-MBA will infiltrate the fast track third level, it is very common for a second level developmental employee to have an MBA. In fact, there are more MBAs running around than there are positions that really require the specialized training they bring to the party.

Consequently, few MBAs get the opportunity to immediately use the skills they have learned in graduate school. Most new MBAs rush around the employment scene shouting, "I'm going to save your company! I've got my MBA!" and employers whisper back: "Oh no you won't! You won't even get close to having any bottom line authority until you show us what stuff you are made of." Most of those MBAs end up in the second level with a good career path in front of them, which, given some luck and a lot of hard work, will eventually give them a position of responsibility where they can use all those skills they have worked so hard to master.

If you have earned an MBA, your attitude about that degree should be of great interest to your interviewer. If you demonstrate that you are egotistical, that you feel you are due more money and status because you have that piece of paper, you will be shut down like a leaking nuclear power plant. If you have an MBA from the graduate school extension on your military base, accept that for what it is. If you have received your advanced degree from Barely Accredited University, which advertises in the classified ads in comic books and on the back of matchbooks, don't **ever** confuse that degree with one from Stanford, Harvard, Wharton, etc. Those top MBA schools produce a special kind of animal and to set yourself up in competition with them in the interviewing process is to invite almost certain disaster.

Your attitude about any advanced degree you hold should be low key: you don't expect it to do a whole lot for you, you've learned some good things you might be able to use eventually, the time was right to get it. Why low key? Several reasons: first, employers have had extremely mixed success with fresh MBA employees. Most could tell you unbelievable horror stories about overly-exuberant MBA (former) employees. Second, the employer may have only Level Two positions to offer, and if you come across too strongly, the recruiter may think you want a Level Three, and you will be out the door to seek it in someone else's company. Finally, the louder you talk about your MBA, the greater the opportunity for the recruiter to translate that into salary dollars. You might find that you are pricing yourself out of the market without ever mentioning money. Be humble. It'll go a long way.

Top business schools have a lot of status attached to their names, and it will open doors at the third level which are not opened to ordinary folks. That might gripe you, but it's true. If it gripes you enough, go to one of the top business schools and become part of the elite. Military leaders with good undergraduate degrees are prime candidates for graduate schools that like some maturity in their students. If you have not thought about it, perhaps you should. And I offer some help:

First, who you are and what you have set as goals for yourself are important facts to consider in making the decision whether to attend graduate school or not. However, even more important is the amount of knowledge you have gained about the field you want to enter.

One of the reasons people are unhappy in their careers - or in their lives, for that matter - is that they have made a commitment to a specific direction, found out that they made a mistake, but continued on that path because they had so much invested that they couldn't turn back. Certainly you knew some people like that in the military. You will know people like that who are civilians.

If you lock yourself into a graduate specialty without real understanding of that field, you might be committing yourself to a course which could be very difficult to change in the future. It's like the bright student who works hard to get into medical school, only to find out that he throws up at the sight of blood. If you commit yourself to an MBA in Finance, for example, you had better be darned certain that a financial career is what turns you on, that there are employment opportunities in it, and that your understanding of the financial career path corresponds to the **work world** reality of the same. The last is important because the **academic world** version of what you can do and where you can go does not always match that which is real, sometimes in the very way a military recruiter describes what is possible in the military. Got the picture?

The very best situation would have you working in the field of your choice for several years and then making the break to graduate school. Nice scenario, but probably not very realistic. As an alternative, you could go to work and determine the value of an advanced degree while you are employed and then attend a graduate course at night at a reputable university. Most major companies have tuition reimbursement programs so the degree costs you little or nothing. However, it is a real effort to get in gear to go to school after work and on weekends. It takes incredible dedication, but if you are willing and able to do it, it could pay long term dividends.

There are no absolutes in the field of employment. All the above could be true part of the time; none is true all the time. Not all fast trackers have MBAs, some non-developmental

Level Ones become Level Twos, some Level Ones become chief executive officers, and some of those Level Twos become Level Threes. Your **entry** into these levels is in large part dependent upon your educational background, and there are even exceptions there. The purpose of this chapter has been to set a scale in your reference, and to give you some idea of where that scale matches your aspirations and qualifications, long term.

So much has been written by so many people about the piece of paper known as the resume, you have a right to be confused. Everyone, from the corporate president to the newspaper helpful hints columnist is a self-appointed expert. I will state the facts, as I know them from my experience of reviewing literally hundreds of thousands, as briefly and as clearly as possible.

It is most important that you understand that the **purpose** of the resume that you send out to a prospective employer is to **stimulate** interest, so someone will call you up and ask you to come in for an interview. Its purpose is **not** to motivate that person to call you up and ask you to go to work. The candidate who writes an exhaustive life story, including an accounting for every day breath has been drawn, has the wrong idea. You cannot substitute your resume for an interview.

If I am a manager, one thing is a set-in-concrete given: my time is always limited; I put a premium on its value. When I receive your six-page resume, and with it the implied obligation to spend at least fifteen minutes to start to digest all the information, I won't do it. Your resume is tossed in the "thanks, but sorry" basket, which is often round in shape. I simply do not have the time.

But let's say that I do, through some unusual circumstance, have the time and desire to thoroughly review resumes. My company spent $1800 on the advertisement; I should review the products. The chances are good you will have included some information on those six pages that I can use as justification for tossing the resume in the same round basket. It could be something as small as your grade point average (which should never be included on a resume unless it is well above average) or as significant as your having included the information that you were fired from your last job (that sounds a bit dumb, I know, but believe me, I've seen it!).

Even if you are clever enough to avoid including negative information on those six (or less) pages, I still have serious reservations about your judgement. A long resume is not a resume – it's a book, and an expensive one at that! If you look up the word "resume" in your *Webster's*, you will find: "a *short* account of ones career and qualifications...." Need I say more?

Your resume should include all those positive and pertinent facts that will activate the interest of the recruiter who reads it. If a match exists between what you have to offer, and what the recruiter needs, you will get a call for an interview. If no match exists, the most beautiful and artfully crafted resume in the world cannot bridge the gap.

During the interview, you will give details which fill out the picture. Favorable details, certainly, but possibly some which might not appear all that positive in black and white. If you apply the principles you learn in this Handbook, you should be able to sell over any of those unfavorable points. Without a doubt, you will be able to do it better in person, face-to-face, than on paper.

The overwhelming majority of applicants should limit their resume to one page. Even though you have been in the military for several years and have a lot more experience to write about than the fresh college graduate, you should also limit your resume to one page. The longer the text, the less the chance it will be read and even less the chance that it will be read thoroughly.

There is an important exception to the one-page rule above. In cases where the purpose of a resume is not that of generating interest so you will be invited for an interview, more detail may (big word!) be warranted. If, for example, an interview appointment has been set up by a third party such as an agent or friend, the purpose of your resume changes. Now it becomes a vehicle for more detailed communication and a source for questions that will make up your interview. You can use that to your advantage. Your two-page resume has replaced the company's application form, in effect. The importance of that document is discussed in depth in the next section.

What should go on your resume? Because of your military service, you do have some unique experience worth relating. And that is exactly what you should do: stress the achievements you have attained while in the military, **in civilian terms as best you can.** Perhaps you were able to save money or time for the military by being more efficient in your motor pool inventory control. That is something you can tell a business recruiter relative to the objective at hand. In the same manner, increasing morale, motivating unhappy subordinates, carrying out your commander's unpopular policy, getting rid of deadbeats, solving drug problems, and the million other duties that fall to the military leader are translated, albeit with some effort, to language a civilian can understand.

Please, do not use military jargon as, (and I quote) "Maintained operational control of 50 employees. Implemented personnel realignment." If you use that style, you will perpetuate the stereotype that military personnel are stiff and formal, and defeat the whole purpose of relating your military experience to the recruiter in civilian terms.

There are many resume writing services standing ready to serve you. They will compose a resume for you, have it printed and present you with the finished product for a fee. More often than not, the resume will be printed on colored paper (which is more expensive than plain white), be at least two pages (you pay by the page), and will have one career objective (e.g., sales representative) which renders the resume useless if you want to apply for a position in any other field.

There is a way around this expense and I'm going to tell you about it at the risk of receiving all kinds of hate mail from Lifetime Resume Services™, which is affiliated with KCE Publishing:

Write your own resume from a reference book. I have included a couple of sample resumes at the end of this chapter. Use one if you like it; if not, improvise. Despite what "experts" will tell you, there is no one way that is absolutely correct, and the definition in Webster's Dictionary does not include a set format.

However, there are several **facts** that are absolute: your resume must be neat, produced on good quality paper, and contain no spelling errors or factual misrepresentations. Less absolute are the facts that it should be attractively composed and written in the third person singular. Your resume represents you, whether you created it or a service did the job for you.

Before you sit down to your typewriter or word processor, you should consider the need to give the recipient certain basic information. Obviously, you will want to give contact information: address and telephone number where you can be reached. If you have the luxury of access to an alternative number, give both with the note that one is a message number. You may also want to include the date you will be available to go to work. There will be high interest in your education and any training that you may have had which is applicable to the job. In most all cases, there will be interest in your career objectives. And this is where the professional resume writers make their money. If you have just spent big bucks for a professional resume with one objective on it, what happens when you want to interview for a different one? Buy another resume?

Maybe you'd think that you should have a broad enough objective to encompass many different jobs, so you could use the same resume for many different applications. That would be wrong. Broad objectives that take in every position from janitor to vice-president of marketing read as if the applicant is interested in any job, not a directed career. The message is; "I'll do anything", and if I, the recruiter, am looking for a particular individual to fill a challenging position in advertising, sales, production management, finance or whatever, I'm going to be most interested in someone who knows what they want, and is willing to state it directly. I don't want to have to interpret an objective and decide whether or not my open position would be interesting to you. If I have an opening for a financial trainee, my interest level will be much greater in the resume announcing "entry level position in finance" than in the vague "position with responsibility leading to management," even if the qualifications are identical.

The solution to the problem of someone who has a variety of career interests is to have more than one resume, each with a different objective to fit the opening at hand. But at $50-$60 a pop, that is very expensive.

So I recommend this: create your basic resume on a high quality electric typewriter or a computer programmed to a laserprinter. It will be helpful if you have continued access to this equipment while conducting your job search. You will notice that I have omitted reference to a dot-matrix printer. Even the most expensive dot-matrix has a tacky end product which cannot compete in this arena. Your resume should look outstanding.

What you do is change the objective to fit the need. With an electric typewriter, the best method is to use white correction fluid or paste-over paper carefully. Obviously, with a computer to laserprinter setup, the job is easier and looks better. Take your newly-created master copy to the printer and have copies made. Some will do it while you wait. For smaller volume, there are photocopiers (which are possibly available in the same printshop) that make excellent copies and can do so on bond paper you have for the purpose. The changed objective will not be evident on the photocopies. You will not have paid for 100 copies of a resume when you need only ten. You will be able to change information as needed (such as your telephone number or objective) and you can do all of this at a fraction of the cost of the professional resume services. Don't ever say I didn't help you out.

Now let me put in a perfectly blatant plug for my friends at Lifetime Resume Services™. If you do not have the equipment, time, energy or skill to compose and produce your own professional resume, this service will do it for you for the rest of your career at a very reasonable cost. There is a one-time charge for creating the document, which is stored on disc, and small fees for updating and expanding your resume as your career grows and your jobs change. With facsimile (Fax) technology and overnight mail you can have an updated supply almost immediately, and no need to futz around with the mechanics. Write to us at the address in the front of this book and we'll gladly forward your information request to Lifetime™.

Some last words of caution. Check and recheck and have someone else recheck your resume for spelling and typo errors. Many resumes have obvious mistakes that immediately jump out at anyone with an eye for detail. It is possible that they are typographical errors made by a disinterested typist. It doesn't matter. You are accountable nonetheless.

Your resume represents you. It can generate interest in your background. It can add validity to the fact that you are a thorough and well-organized individual. It can represent you after you leave your interviewer. It can reflect your attention to detail and flair for style. But it cannot take your place in an interview.

Sample Resume
Samuel B. Cracker, Jr.
1404 Long Road, Visalia, TX 79999
(817) 999-9999

Education:
  B.S. Animal Husbandry 1982
  North Carolina Agriculture & Technical State University
  Greensboro, North Carolina
  50% financed by part-time employment; active in extra-curricular activities.

Experience:
August 1982 to present: United States Army, Lieutenant

November 1987 - present, Division Assistant Fire Support Coordinator. Responsible for planning, coordinating and integrating fire support at the division main command post. Developed and executed a detailed training program which integrated six Aerial Fire Support Officers into the division's war-fighting capability. Contributed to the development of two key fire support documents: a fire support handbook and a fire support battlebook. These two documents are now used as source documents for development of doctrine at the field artillery school.

April 1986 - November 1987, Division Logistics Plans Officer. Responsible for all division-level logistics operations. Planned the logistical requirements for REFORGER 87, one of the division's largest exercises since World War II, involving movement to West Germany and return to the U.S. Performed frequent briefings on logistical status to the Division Command Group. Supervised four personnel.

December 1984 - April 1986, Platoon Leader. Advised battalion staff on logistical issues for tactical or strategic actions. Prepared supply, maintenance and ammunition reports. Additionally, performed duties as the battalion recreational officer organizing all sports-related activities.

Military Awards:
Army Commendation Medal (3), Army Achievement Medal, Humanatarian Service Medal, Army Service Ribbon, Overseas Service Ribbon, Senior Parachutist Badge.

References: Available upon request

Date Available: June 1, 1989

SAMPLE RESUME

NAME: John W. Logan

ADDRESS: 327 N. Elm Avenue, Apt. #3
                Littletown, NJ 01403

TELEPHONE: (214) 888-8888 message: (214) 999-9999

OBJECTIVE: Sales Leading to Management

LOCATION: Open

AVAILABLE: For Interviews: Now
                For Employment: July 1

PERSONAL: Age: 29; 5'7", 165 lbs, Married

EDUCATION: B.A. Business Administration 1982
                University of Littletown
                Littletown, New Jersey

Activities: High School (optional): Varsity Football, 3 years. Elected Captain in senior year, MVP Award, Honor Roll student five semesters. College: Football Scholarship, Paley Award for outstanding scholarship, Football, 4 years, Fraternity officer and Rush Chairman, Member of Business Association.
Note: Financed 50% of college expenses by working as inside sales rep part-time and during summers for Acme Manufacturing.

EXPERIENCE:
June 1982 to present: Officer, U.S. Army. Captain, Company Commander with supervisory responsibility for 117 men and 16 women in a logistical and inventory control unit in Europe. Reduced time necessary to fill orders by 35% by redesigning shelving space. Held command of this unit for 26 months, (eight longer than normal) due to outstanding performance. As Lieutenant, aided staff officers in organizing and presenting ideas for change in military procedures to Commanding General. Received two military decorations for achievement.

(Note the absence of first person singular. Traditional, but not absolutely necessary for a resume. Also, we are pressed for space here; normally, you can use white space to balance the page.)

My, isn't this getting down to the basics? If I had never seen what an applicant could do to an application form, I'd never have thought to include this section in our Handbook. But I have, and so I will.

I have a strong feeling that a recruiter has the right to believe that if a candidate fills out an application form incompletely and in a sloppy manner, the applicant's interest in the recruiter's company is less than 100%. The amount of effort you invest in that piece of paper is a direct reflection of your interest level. And if you are not very interested in that company, why should that company be interested in you? Your application form is a **direct** indicator of your interest level. Your interviewer will evaluate you based on its appearance as well as its content.

Additionally, it is the first opportunity the recruiter will have to evaluate how you can complete a task: filling out the form. While resumes are important, applications are a much better indication of your attention to detail. You cannot pay someone to fill out an application form as easily as you can pay someone to build a resume for you. How you perform this task is more important than it may initially appear.

There are at least four times when you could be given an application to complete:

    a) a day or so before the interview,

    b) after your interview, to be completed and mailed later,

    c) forty or so minutes early, to be completed prior to your interview

    d) after the interview, to be completed before you depart.

In case a) and b), you should be able to build the best application and there are no excuses why you shouldn't. If you have no instructions to the contrary, type the application form if possible, but only if you can type without making lots of mistakes or have a correctable typewriter. Some applications are required to be filled out by hand rather than typed. If you can bring the form home with you and fill it out at your convenience, you can take these steps to assure that it looks terrific:

First, make a copy at your local copy machine. This will be an inexpensive insurance policy. Using the copy, complete all the information requested.

Second, proof read, or have someone else proofread, the entire completed copy to check for spelling and grammar. Absolutely nothing is as big a turnoff as an applicant who lists "bussines management" as an objective.

After you have proofed your copy, transfer the information onto the original form, taking extra care for neatness.

Here are some tips:

1. Where you are asked for names and addresses of people or organizations, give full information - name, title, address with zip code, and telephone number including area code.

2. Avoid folding an application form. Go to the extra trouble of buying some 9" x 12" envelopes, even with thick protectors, and mail them flat. A skrunched-up application will not work to your advantage.

3. Some questions on applications are actually interview questions. An example would be a blank space which asks for "anything else you can tell us which might help us consider your application." You will recognize that as an open-ended interview question later in this Handbook. Most candidates leave it blank or write "regional Tiddly-Winks Champion 1987." You will use the space available to answer this type of question with depth, giving as much positive information as possible.

4. For reasons we will discuss later in this Handbook, salary requirements should read "OPEN," unless you have definite information to the contrary.

All these things are easier to accomplish when you have an hour to prepare the application form at home. But what about cases c) and d), when you are given the application to be completed on the spot?

A little foresight will pay large dividends in that case. If you look at 25 different application forms, you will see that they all have certain things in common: your work history and education, your references, and other personal data. While you are in the process of your job search, it makes sense to have this information with you at all times. The information can easily fit on two or three 3" x 5" cards you can carry with you. Here are some of the details that should be on these cards:

Your military history. Include your assignment dates, ranks, titles, salaries (don't include benefits), numbers of personnel supervised, special schools attended, and dollar value of equipment you were responsible for.

Your civilian work history. Include every job you have held, the names of supervisors, the address of the business, including zip code, your starting and ending pay rates, the dates of employment and your job title. If this list starts to get longer than six jobs, make certain your list includes relevant experience. Most forms have the guidelines of six to ten years or four jobs, but make sure you include other positions outside those guidelines that are relevant to the position at hand.

Your education. Include the name and location of each school you have attended, including your high school, the major field of study in college, the dates you attended, your degrees and your grade point average, both in your major and overall. You might also add a complete list of extra-curricular involvement in high school and college, as well as civic or other appropriate activities.

Your references. Include at least eight: three people you know very well, three people you have worked with (not for), and two people you have worked for (who know your work habits).

Job seekers tend to get overly-involved in the subject of references, and there is really no reason to be very concerned. Don't feel that your references need to include five bank presidents, three oil barons and an international tycoon. It's nice to list influential people, but you gain nothing if these people barely know your name. Also, who ever said that references had to be older than you? What about that close friend you have had for the past ten years, or the guy you worked with at the college bookstore? What you really need are eight people who know you well, and will say nice things about you and your work habits, character and personality in the unlikely event that they are called by a prospective employer.

If you know someone who actually works for the company you are applying to, and are certain that person would say good things about you, make sure that he or she is listed as a reference. Information from internal sources is given higher credibility than external ones.

List these people, their complete current addresses, telephone numbers, titles and occupations. Then, sit down and write a short note to each of them telling them that you are on the job market and are using them for a reference. This prepares each person for a phone call or letter from an employer, and joggles their memory if, as in the case of a past supervisor, it has been several years since you have had contact. Who knows, perhaps they will call you with a lead!

Do it! Put this information on those 3" x 5" cards. This is the kind of idea that could make the difference between success and failure in your job search.

I would like to add two small but important thoughts to this section.

If you have a spelling problem, invest a couple of bucks in a small pocket dictionary and make it a part of your basic equipment load for the duration of your job search. Poor spelling is an enormous handicap. If you are given an application form or other written assignment, spend any extra time you have looking up words you doubt. Since you have all the information at your fingertips, you should speed through any application form and have time left over.

Obviously, if you are handed a form to complete and return on the spot, you will not be able to type it unless the secretary/receptionist is open to negotiation. If you have neat penmanship, no problem. But if you are like most people and have poor to passable handwriting and printing, you will want to be extra careful. Here is a suggestion that worked for me. In the military, I always wrote with a government-issue ballpoint pen, click, click. My handwriting was legible, but not something you'd want to use as an example in a penmanship book. Someone convinced me to try an ink pen (back then it was the type you filled from a bottle and always ran out of ink at the wrong time). My penmanship improved 100% because the fountain pen does not slip around on the paper as much as a ballpoint, and now it is the only writing tool I use when doing something that needs to look great. If you have sloppy writing, give it a try, or experiment with some of the non-ballpoint pens that are available and give the same effect.

Why have we spent all this time emphasizing the importance of application forms when most books spend all their emphasis on resumes? Besides being interpreted as a direct indicator of your interest level and an opportunity for the interviewer to evaluate how well you perform a task, this same application form will become a **permanent** part of your personnel file when you are hired. Every time you come up for consideration for promotion internally, or are interviewed by internal managers, up pops this form from your file, even if you filled it out 15 years earlier! It's true! This long-range, potentially positive or negative impact it carries is more than enough reason to pay special attention to this deceptively simple task.

I want you to be honest with yourself for the next few moments while you answer this question: when you were reading the last section and learned that I thought you should sit down and list the information necessary to complete an application form on 3" x 5" cards, what was your immediate reaction? *Be honest!!*

Was it, "Hey, that's a good idea. As soon as I'm finished reading, I'm going to do that!" Or was it a strong enough reaction to make you stop and do it before reading further? Maybe you marked the page to remind yourself. If it was that kind of reaction, fine, keep on reading, and make sure that, sooner or later, you fill out those cards and keep them with you.

BUT, if your reaction was, "I don't have to do that," or "What a bunch of ......" (fill in your own word), let me make a suggestion. Stop reading, go find an envelope and return the Handbook to me so I can refund your money.

If you are not motivated to learn and apply these principles and ideas, you have wasted your time and money in purchasing this Handbook and reading it this far. Let me assure you the effort is all uphill from here; we've just started the basics. Millions of people waste time and money on all kinds of books and programs of self-improvement. All these books and devices are a waste of money if you are looking for an **effortless** solution to your problems and needs. So go ahead, if you don't want to make yourself competitive in the interviewing process by working at it, send the book back to me if you bought it from KCE Publishing, or return it to the bookstore so they can send it back through distribution channels.

If you **are** motivated to do the things outlined (and please note that they are not really difficult to do, but require honesty and commitment), I'd like you to keep this thought in mind as you continue: **the techniques outlined in this Handbook can really make you more competitive in the employment market, if you are willing to perform them.** The overwhelming majority of people who interview for their first full-time career positions have spent so little focused time in preparation, they cannot even discuss three of their positive personal qualities without stammering, searching and delaying, trying to think while talking. Believe me, I've seen it, all recruiters have seen it, and it's true. With a little effort and this Handbook, you will be so much better prepared that you can make even more qualified competition look pale by comparison. Believe me.

In addition to the time you have spent in college, you have invested four to six years, perhaps more, preparing yourself for civilian employment while in the military. That preparation will not be fully evident unless you invest a little more time in specific preparation to confront the employment market.

So, if your attitude is right, let's continue.

"May I first apologize for any time of yours I may have wasted in directing this correspondence to you."

"My low GPA is not indicative of my true ability, learning speed, or my sharp, analytical mind."

Isn't that frightening? Can you believe it? Yet those are two real examples of statements in cover letters that I have received in the mail. Let's spend a few minutes together on the subject.

Cover letters are sent to prospective employers along with your resume in response to a lead or as part of your mailout campaign. The only reason you need a letter is to introduce yourself and communicate some pertinent information that is not on your resume. An example is to announce your intention to travel to Chicago in anticipation of interviews with several companies.

Make your cover letter sweet, short and to the point.

Dear Mr. Jones:

I expect to be in Chicago in three weeks and would appreciate the opportunity to discuss sales positions at Acme. I was a part-time clerk for your plant in Omaha during my junior year in college (as you can see from my resume), and feel that Acme is a company I'd like to know more about now that I am seeking a full-time career position.

I will call your office next week to see if we can arrange an appointment.

    Sincerely,

    Sidney Signup

Control your urge to make a cover letter into the interview your resume is not. Busy recruiters won't read long cover letters any more than they'll read long resumes.

A few words of advice: type, or have laserprinted, all of your cover letters as originals. Do not reproduce bodies of the letter at the printer and then type the name and address of the recruiter in the heading. Adding the name of the company in the blank space you so cleverly left at the end of a line is always obvious. To the recipient, it means you have massed-produced a letter which would have had more impact if tailored to the specific company or opportunity. Indeed, if you are sending your cover letter and resume in response to a specific opportunity, the time and effort it takes to write a special edition of your cover letter emphasizing how your experience matches the requirements of the position will be time well spent.

If you have an enormous volume of cover letters to send (as in a mass mailout), invest in secretarial services which can access a computer or automatic typewriter generated format. The "print merge" command on a first class word processing program will merge text and lists of recipients. It will be expensive, but we are talking about form and first impression here!

Another use for a simple cover letter:

Sir / Madam*:

I am aware that most companies keep unsolicited resumes for a limited time, so I am forwarding another copy for your use. I would like to hear from you if a position becomes available for which I might be considered.

S. Signup

---

* If you don't know whether the recipient is male or female, you'd be well-advised to use this sort of greeting in *all* your employment correspondence.

As you can see from the two real examples at the beginning of this section, a poor choice of words or unnecessary discussion can easily turn your cover letter from an asset into a liability or even a source of comic relief. Try writing a couple of different versions and get some advice on which sounds the best. Remember, brevity is the key, but avoid the extreme –you don't want to sound terse.

It would be fair to say that a major goal at this point in your life is to create options: a number of opportunities in the civilian sector from which you can choose a particular one as the best possible situation for you. This may be one of the more important and difficult decisions you will ever make, and to jump at the first option offered to you could be a mistake.

Most people on the housing market look at a number of homes before they choose the one they want to buy. They know all they can about the structure, price, tax rate, neighborhood, appreciation potential and school district before they make a commitment. Once they have signed on the dotted line they know there is little chance of turning back without severe financial consequences. While you might not immediately lose a good deal of money if you make the wrong career position decision, or take an offer where there are unanswered questions, the effect is the same as if you were buying a home with poor prior research: wasted time, money and effort. Just as there are necessary improvements in the house you buy even after careful consideration, so are there facts which you did not know or could not know about that will appear only after you have accepted a position. But with a certain amount of care, these factors will be minor, and will not affect the "livability" of the job.

I have chosen the parallel between buying a house and accepting a career position because there are distinct similarities. The choice in either case has emotional overtones and long-term commitment. Once you have found the house or position you want, there is the strong temptation to own it **now**, before someone else sees the positives the way you see them and beats you to the punch. And just as easily, one could be influenced by the temptation to continue looking, in hopes that the next house or the next position will be just a little better than the one you are looking at now. And the search goes on, and on, and on.

Either extreme reaction, grabbing the position or looking indefinitely, can be self-defeating and result in an unhappy situation. There is no "perfect position" any more than there is a "perfect house." You must resist either extreme temptation and make a good decision in a reasonable amount of time. The smart method is to investigate a number of positions and, considering your individual time restrictions, isolate those which best fit your needs and decide from among those options. What is important is that you decide for yourself. It is your decision, and only yours. You will have to live with it long after anyone who has given you advice has forgotten about it and you.

Sounds neat and simple, doesn't it? But how do you go about interviewing all these companies at once? Don't they expect you to give them an answer? Read on...

Consider the word "offer." It carries with it a favorable connotation. An offer can be accepted or rejected, and it definitely leaves room for negotiation after it has been considered. If you approach an organization saying "I want to go to work for your company," you are committing yourself with only the barest of facts in your knowledge pool, and you may find that the job has holes in the roof and termites in the cellar. There is an appropriate time and place to make that commitment, but that time is certainly not in the first stages of exposure to the company or position.

The word "offer" has less of a commitment. If you go to an organization and tell them you are interested in an *offer*, you are saying you wish to know the facts - more about the company, more about the job, and more about the other important details. Once you have all the information, you make your decision. Some candidates have difficulties accepting this as a way of operating, but let me assure you that the companies you talk with will have a similar attitude: what have **you** got to offer **us** in terms of education, leadership and ability? **Then** they will make up their minds about whether to give **you** an invitation to join or a rejection letter.

The proper time to evaluate a position is when you have all the facts about it. It is when you have an offer, not when a friend who has interviewed with the same company tells you it isn't any good. This is a very personal decision, and what excites one person may easily leave another cold. If nothing else, certainly you have learned that in the military.

You are entering the civilian job market with a number of prejudices about different kinds of jobs, about what you want in a position, and about what kind of company you want to work for. It is natural to have these kinds of pre-conceptions. It is also natural for people to change their ideals when they find that they are faced with a group of opportunities which only has remote resemblance to what they had in mind originally—sort of like the couple who settles for a two-bedroom cottage rather than the mansion on a hill they pictured as an ideal. Ideals are fine, but you must eventually come home to what is real. Make your decision on a particular position only when you have an offer to go to work, not in the first interview, or after you have read the promotional literature. If you reject a situation because it is not within the boundaries of your ideal, you might be looking back at the passed opportunity with different feelings when you find your real offers are not anywhere near what you expected. It is almost impossible to go back and open closed doors.

Your objective at this point is to get offers. You should make it a policy to talk with any potential employers who will talk to you, whether the situation is initially attractive to you or not. Stop worrying about wasting their time! Interviewing lots of people is part of the recruiting process. If a recruiter were to hire everyone who was given an offer, the company would be overflowing with people in a very short time.

To be sure you have an offer, you must be able to apply all these points of definition to your situation:

- what you will be doing

- location where you will be doing it, and what expenses will be paid to move you there

- when you will start doing it

- how much in salary and benefits you will be earning initially, and

67

• who you will report to

With the possible, but never recommended, exception of the last point, you cannot make a decision (and do not have an offer!) unless all these points are covered. Additionally, and not necessarily minor, considerations (such as who pays for the incidental relocation costs) should be cleared up at the time the formal offer is presented or shortly thereafter, as you consider the whole picture and need to focus on the details.

Other points: you do not have an offer until you have one formally. That means when a representative of the company specifically asks you to go to work and can answer all the definition points listed above. Someone saying that they think they will be making an offer this Thursday, an employment counselor assuring that you will receive an offer, or a recruiter saying that the hiring manager will be making an offer soon are all good examples of *not* having an offer. Unfortunately, things do not always work out the way people expect. If you take their word as truth and turn down or turn off other opportunities, you could easily be starting your entire search all over again if something unexpected (hiring freeze, internal promotion, better candidate) happens.

APPEARANCE AND FIRST IMPRESSIONS

I HOPE YOU HAVE THE ABILITY TO READ THROUGH THIS ENTIRE SECTION BEFORE MAKING A JUDGEMENT ABOUT THE CONTENTS.

Almost everyone involved in employee selection will tell you it shouldn't be done, that it isn't right or fair, but that it exists in practice. I'm referring, of course, to the practice of evaluating an applicant based on first impressions, and giving that impression a heavy weight in the decision to pursue or reject the candidate.

In the same way your application form is an indicator of your enthusiasm and interest, so too is the amount of care and effort you spend on your physical presentation. With justification, a recruiter thinks that you are at your very best when you are interviewing. It is a given that you will change your appearance once you are hired. Your dress will probably be less formal, your hair will get longer or be in a less formal style. Sometimes people hired into the company are unrecognizable six months later. Who is that guy?

Even if you know the organization does not have a specific dress code, or has one which includes a standard of Levis, no ties, and tennies, you should **never** confuse your appearance standards with those of the employees. They **have** their jobs, you are looking for yours. This may require that you homogenize yourself for a while in order to conform to a very unspoken and unwritten standard. I hope you can handle it. There will be plenty of time later in your career to make personal fashion statements.

Some people have the ability to look sharp with very little effort. The rest of us have to work hard at it, and the intent of the following is to assist you in that effort.

Not many people are totally reactive and negative about longer hair on the male human animal. Even the most conservative have accepted the fact that you are not necessarily out to overthrow the government just because your hair touches your collar. But, since you do not know your future interviewers' tastes on the subject, it would make sense to go the more conservative route and cut it shorter than your favorite rock star's. No, not sidewalls. But as short as will allow you to be comfortable.

If you are still in the military, you may be uncomfortable with letting your hair **grow**. Habits change slowly, but if you think you look smashing with sidewalls, you don't.

Most of the people making decisions on your future employment will be much older than you. Many of these persons will have values and standards very different from yours in many ways, and hair length is just one of them. Interestingly enough, many of them will have never been in the military, and carry all sorts of misconceptions about military people, and they may react to your overly-short hair. Have a good barber/stylist cut it so it looks neat. If you have been letting your roommate or spouse cut your hair, now is the time to spend some money and have it done professionally. Both men and women can easily improve their overall appearance with a little extra effort in this department.

What should you wear? This could depend upon what you have in your closet if you are going to interview tomorrow or the day after, or on your financial status if you are fortunate enough to have the time to improve your wardrobe before your interviews begin.

For both men and women, my initial advice is to avoid the extremes. This comment may seem obvious, but I remember interviewing a young man wearing a purple jumpsuit and jogging shoes. I also remember (quite fondly, actually) interviewing a young woman who was braless and in a tight T-shirt and Levis. I'm not sure what these individuals had in mind when they got dressed the morning of their interviews, but one thing that did not impress me was their seriousness about working in a managerial environment.

For women, a stylish suit is entirely acceptable. Outfits should be chosen with business in mind; not too flashy, but not too conservative either. You may find that a scarf or other conservative ornament will add to the businesslike quality you seek. AVOID jingling bracelets which raise a racket every time you move, flashy jewelry, and excessive use of perfume. Most interviewing takes place in a relatively small office at close quarters and if you just doused yourself with Essence of Sweetpea Extract, you might just asphyxiate whomever is interviewing you. Men, you take note, too. Some aftershaves and colognes would make a buzzard wretch in a closed room. Again, as in the case of hair style, do what you can live with, but be realistic. Significant career positions that allow you to dress as you do at home on a Saturday afternoon are few and far between.

Men, you will be doing yourself a grave disservice if you show up for any interview in less than a coat and tie, and there is a strong case for always interviewing in a suit. Remember that a recruiter is thinking you look as good as you are ever going to look. I would strongly recommend that you commit some money (at least $400) for the purchase of your interviewing wardrobe, and buy a good-looking, up-to-date suit for that purpose. Military leaders often have had to depend upon dress uniforms for formal, work-related social events, casual clothes for less-formal events, and uniforms for their workday attire, and neglect their civilian suit wardrobe as a result. The time has come to do something about it and the stakes are too big to depend on that four-year-old brown outfit or suit you ordered from the Hong Kong tailor at the exchange.

Your purchases should also include:

At least three good shirts or blouses (long sleeve, not short) of a solid color that will go with your suit.

At least two coordinated accessories, ties or scarfs.

Some formal shoes. You cannot expect to wear those broken-in loafers to an interview, comfortable as they are. And, for the men, buy some calf-length hose, so that when you sit down three inches of your hairy leg doesn't show.

Does all that sound like a bit too much for you to stomach? OK, don't do all of it. Just the psychological lift you can get from looking good can give you the confidence to do better in an interview, so I think it is a good investment. Whatever you decide to do, understand that these are all things I notice when I evaluate an applicant, and if I notice these things, you can be assured that other recruiters do as well. As a matter of fact, let me add a couple of things, if you are still with me.

You can detract from your whole business image if:

You wear a sports watch that may allow you to read the altitude of your airplane or the depth of your SCUBA dive since that is rarely information one needs in an interview. Or a watch with a tooled leather band your father-in-law made in leather craft shop.

You wear several rings on either or both of your hands. If you are a ring freak, limit your indulgence to one on either hand.

You wear a fraternity or sorority pin on your lapel or collar (come on, that's for college kids, you are in competition for a position which calls for **maturity**).

You wear clothing which does not fit. If you have had a recent weight loss or gain, make sure you buy that new suit or outfit and make sure it fits, or get your old serviceable one tailored.

And, even a step further, Look at your hands, yes, right now. If you were looking at those hands as an employer, let's say of a hospital or dental supply company, would you be impressed? Would those hands meet the approval of a surgeon or dentist? Or do they look like you have been tearing apart a '75 Chevy engine without a wrench?

All the above is in an attempt to get you to do something very few people ever do seriously: look at yourselves as others, in this case employers, see you. What can you improve?

We have discussed appearance at length, and how it relates to first impressions, but there are still other areas of discussion you must suffer through that add to or detract from that important first impression.

Handshake: play this one by ear. You may find that the interviewer has no interest in shaking hands with everyone he talks with, and will simply ask you to take a seat. If you do shake hands, make sure that it is firm, and neither bone-crushing nor a weak, hand-me-a-dead-fish type. A minor point, but these points have a way of adding up quickly.

"Taking a seat" reminds me of a story. I did exactly as outlined above with a candidate once; stood up, walked to the door, shook hands and asked the person to take a seat. He did, only he walked around behind my desk and took **my** seat. Honest to God! Looked like he'd just moved in and taken over. Well, I think I did a pretty good job of getting us both out of that one with a minimum of embarrassment, but it illustrates well the initial tension of an interviewing session.

Voice: the moment you open your mouth to say "Hello," you must be conscious of your voice projection and how you sound. Your voice is a tool and you can use it to help emphasize a point. It has tonal qualities that can make your audience believe what you say is true, or it can bore with its monotone and literally put your audience to sleep. What do **you** sound like? Have you ever recorded your voice casually and listened to it? Try this: use a tape recorder to tape your voice during a telephone conversation, the longer the better, so that you forget about the recorder. Then, play it back and listen to the inflections. Whew!

Eye-to-eye contact: my personal sense is that the single most distracting thing an applicant can do in an interview is to talk to me while looking out my window, or at the plants, or at the pictures on the wall, or anywhere but at me. It is downright rude, and gives me a feeling that this applicant is either super-nervous or is not telling the total truth, or both. Eyeball that recruiter! You should be talking to that person, not the plants! If you have difficulties with this, train on everyone you meet and talk with. Don't wait until you are in an interview to develop this practice. Look them in the eye. You can do it without staring at the person with a little experience. If you must, draw a bead on the person's nose or chin and look at that point as you talk.

Smoking: don't. Even if an ashtray is present, even if the interviewer is smoking (which should not be the case, out of common courtesy), don't. It shows nervousness and is really a discourteous and negative thing to do. This is particularly the case if your interviewer is a non-smoker or has (worse possible case) recently given up smoking. With all the publicity non-smoker's rights has been getting, it does not take much thought to see why you shouldn't smoke in an interview setting.

Chewing gum: good Lord!

Sitting in the chair: whether in the interviewer's office or in the lobby waiting for your appointment, you should be conscious of the image you are presenting. A person who slouches on the couch or chair, feet out in front blocking the aisle, arms folded behind the head, looks like a slob to anyone. Give the impression that you are ready to interview and are interested in what is going on. There might be some pertinent literature to read, information on the organization. Read it, not the "Sports Illustrated," Out in the lobby you may be observed by the person(s) who will be interviewing you, and not realize that the person who walks by and gives you a glance will soon be sitting across the desk from you.

Men, when you sit down, unbutton your suit jacket so that it falls open and then button it again when you stand up to leave. Reason: you look more comfortable and neat. At home, put a chair in front of a mirror and sit in it, first with your coat buttoned, then unbuttoned. See what I mean?

Notes: don't take notes in an interview. It will distract both you and the interviewer and you cannot possibly take notes and talk at the same time. What would you possibly want to write down, anyway?

Vocabulary: many young people who have become comfortable in the relaxed atmosphere of the fraternity/sorority house, at home, on board ship or in the BOQ, are often incapable of carrying on a serious and intelligent conversation without extreme self-discipline. A good example would be the person who uses "you know" every third word in a sentence, as in: "I am, you know, interested, you know, in the, you know, like the opportunities your company has, you know, to offer." Sounds

74

like they're on drugs, to me. Do **you know** what I mean? This lack of ability to express oneself intelligently **reeks** of immaturity, although I have observed the symptoms in men and women who were supposedly mature and well-educated adults. Equally distracting is the constant use of "OK," "uh huh" instead of "yes," and slang words like "groovy," "shine on" and others you will be able to think of which are much more up to date with the local crowd that hangs out at the mall. Cool it, or the main man is going to do a number on your head, instead, dig it? Totally.

Under **no** circumstances are you allowed to use profanity, no matter how low your interviewer sinks in the choice of words. Period.

Finally, and how this relates to initial impressions will become evident later, get yourself some credit cards with as large a credit line as they'll allow. And don't forget a telephone card as well. Telephone companies once allowed you to charge calls to your home number, but it is an inconvenient and time wasting process now. If you must, bribe your parents or another responsible party and get their card numbers so you can make long distance calls where necessary without a pocket full of change. Promise to pay them for the charges you incur and then do it. After all, your word is supposed to be a bond, right?

I would like to emphasize, now that you have read through this entire section (are you still with me?), that these points on appearance and first impressions may be unnecessary. Indeed, you may be totally justified if your attitude is, "I wouldn't work for a company that judged me on those points, even if they came to me on their knees." If all interviewers had the capability to look past that initial impression, past it to the qualities the individual has inside, you would be able to interview based solely on your credentials. But that is not the case and never will be, because recruiters are humans. And so you must take your appearance into consideration and prepare for improving the image you project.

Can you afford to take the chance that you will not fit into the picture, just because you failed to pay proper attention to your first impression image?

I have such strong feelings about this subject that I believe it deserves a special section in this Handbook.

As you get into the interviewing scene, you will find that many recruiters and interviewers get behind in their schedules and that you will have to wait to interview. Maybe that is impolite, or does not show good organizational ability on their part, but that is the way it is. Now I'm going to tell you that there is a real double standard on this point. It is all right for the recruiter to be late, but you'd better never be. Ever.

There is simply no excuse that can undo the damage. If you know that you should be in a certain place at a certain time, and you are not, you show, or give the impression, that there was something more important going on elsewhere, that you are impolite, that you are not well-organized, and that you are less than 100% excited and enthusiastic about going to work for that organization. You have stepped into the interview with a negative field in front of you, to say nothing about the loss of precious time you could have spent selling yourself to the recruiter.

If you must drive across town for an interview, give yourself some extra time in the event there is a traffic problem, because there will be. If it is raining, allow some extra time. If you have to find a building unfamiliar to you, allow some time, and leave early. If you are going by plane, count on the fact that there will be delays this time, even though you have made the flight fifteen times without delay in other instances. In all cases, you should have the address and telephone number of the person with whom you are meeting. If some tragedy does occur, call that person and explain what the problem is; don't just show up 20 minutes late. If you don't have change for the pay phone, use your credit card. There is no excuse for not calling. You can even call from some airplanes.

There are a couple of advantages to arriving early for your interview (I'd say no more than 15 minutes early, though). First, you will not have to make your entrance into the office with a red face and perspiration streaming as you race to get there on time. Second, the odds are good the receptionist will announce your presence, even if you are early for your appointment, and you might get to spend that extra time with the interviewer.

But most of all, you will show that you are a reliable person. And, if you are late for your interview, all the talking you may do on the subject will not convince the interviewer you have the ability to be in the right place at the right time.

What's the value in presenting documents of reference in an interview? Little. But I'll bet you compile some anyway.

Your collection of paper **could** be useful to you, and you should have it organized in a manner that allows you to find a particular document (e.g., college transcripts) without shuffling through the entire stack. There are several convenient methods of doing this: a file folder, a small ring binder, a folder of the kind used to contain term papers or reports, or a plastic folder with a removable bar to hold the papers in place. I would not recommend a large envelope, as this means that you must remove the entire contents and dig through it to find what you want. Of course, you will not want this folder to contain the original documents, so make crisp copies for your day-to-day interviewing.

You may bring this collection into the interview with you, but I do not recommend that you offer the material unless you are asked for it specifically or feel very strongly that something in it will improve your case. Make sure that it is placed at your side, on the floor and out of the way and not in your lap where it can be an outlet for your nervousness.

It is my opinion that letters of recommendation from military or, for that matter, civilian employers are of little value. I have yet to see one that contains anything less than positive, glowing statements. Since they are all uniformly positive, none has any real value.* The interviewer has no way of identifying with the person who wrote the letter and often the letter is commending you for a job done in an entirely different field of work. It is not a total waste of time to collect them.

---

* I just flashed on a scene where Johnny gives a recruiter a letter of recommendation which says, "Johnny has been a loyal and trustworthy employee during his tenure with Acme Ballbearing Works. Unfortunately, he is a dork and if you hire him, you will experience frustration and expense beyond any value he could ever possibly bring to your or any other profit-motivated organization."

You might run across a particular interviewer who places more value on them than others, depending upon the kind of position you are applying for, and what kind of experience you have had. But in general, they detract from, rather than add to, your presentation. The flow of the interview is disrupted by the interviewer reading this material. Your time is spent much more effectively with verbal communication.

At a minimum, you should have a copy of your college transcripts on hand so you can provide copies for employers who require them. Some companies might require "official" transcripts from your college registrar, but you will find that most companies will take your word that you have graduated from Wherever University and have no interest in seeing even an unofficial transcript until facts are verified prior to employment. If it makes you feel better, include your military efficiency reports, citations and letters of military commendation. But please don't force these upon some poor recruiter who is already having enough trouble figuring out what swims in a motor pool. It's nearly impossible for most civilians to relate to the nuances of a military efficiency report.

Q: "Thank you for coming to the interview, John. As you know, there are at least fifteen companies holding interviews on the base today. How did you happen to choose my company to talk to?"

A: "This was the only schedule I could get on."

Reaction: "Three points for honesty, minus ten for judgement!"

The idea is that a small amount of preparation in this situation can go a long way. If that applicant had spent an hour or so in the public library doing some research, he could have answered the question with some knowledge of the product or service of the company, possibly referring to the excellent training program, the competitive products or services, the problems the industry was going through (and the fact that he would find such problems challenging) and many other small but important points.

I'm not saying you should spend six hours in the library before every initial interview. I think that is a waste of precious time. I do feel, however, that going to the *Wall Street Journal Index* and other periodical indices can reveal some very pertinent information in a short amount of time. After the initial interview, you should spend some more time becoming an "expert" on the company, not only to assist you in your interviewing efforts, but for your own personal information if it looks like you are going to be getting an offer and making a decision. Recognize that you might find out some facts that are not complimentary to the organization. When you discover these facts, you should store them. You might find a use for them in the second or third interview.

If you are in a position to do so, subscribe to and read some pertinent periodicals. In the case of business careers, there are hundreds to choose from. I like the *Wall Street Journal* for detail and *Forbes* for more general scope. Get and stay connected to your career specialty. There are trade publications for auditors, engineers, librarians, winemakers, zookeepers and just about every other career group you could imagine.

The person who is well-prepared becomes an applicant who is really looking for a career with the organization, and not one merely sitting in the chair looking for a job.

If you had the luxurious option to interview whenever you want, when should you do it? If I had my choice on picking any month, I'd pick January through February for these reasons:

a)   Few employees are going to quit their jobs during the Christmas season in December. Being unemployed is a worry that does not go well with being jolly. Ho, Ho, Ho.

b)   If someone is to be fired, there are few companies hard-nosed enough to make an employee leave as a Christmas present, right before the year's end.

c)   Employees are often paid bonuses or commissions on an annual or semi-annual basis, January 1 and July 1, and few employees are going to walk away from several hundreds or thousands of dollars.

Consequently, people who would normally leave their jobs wait until after the first of the year. All three of these situations create a flurry of interviewing and hiring for about 45 days after the New Year, so it is often a good time to be on the market and ready to go to work.

However, if you are on Christmas leave and traveling to your home town or to the location of a company with which you would like to interview, make the effort to interview. As we will see later, few companies will refuse to interview a decent candidate on a self-financed interviewing trip. Another time of year that is advantageous to the military candidate is June through September, when college recruiting quotas have not been met and the company needs candidates.

As far as the best time of day is concerned, you should, if at all possible, avoid interviewing in the late afternoon. Again, the time of day is not going to turn gold into lead, but if you and your interviewer are both fresh and eager, it will go better for you. I would think the worst possible time to interview is at 4:00 PM on a Friday. You know how **you** feel on a Friday afternoon, and a recruiter is looking forward to that weekend, too. Use your Fridays to mail resumes or write letters to employers.

In today's competitive market, it is no longer safe to assume that a hiring company will absorb all the costs of relocating new employees. Many of the major ones can, since there is room in their budgets for those expenditures. It is unwise to assume that your future employer will pick up all the costs, however, and it could prove to be more than a little embarrassing (and expensive) if you discover that the company does not pay those expenses after you accept the offer.

Many people starting their first career position out of the military generally have not had the time or money to acquire an enormous amount of household possessions (or "stuff"). Some could move from California to New York in a couple of suitcases. That is the best situation to be in when you are on the job market: very flexible.

I feel, however, that I should address the percentage who have, through various circumstances, acquired furnishings and the related paraphernalia of a house or apartment.

First, I think you should assume the worst possible case: you will be required to pay all costs of moving yourself and your belongings to the new location. This cost factor should be taken into consideration when you are deciding which of your job offers you should accept.

How much do you own? Unless you have done it professionally, it is almost impossible to estimate accurately the cubic footage and weight of those furnishings. So have a professional do it. Call any major moving company and tell them you would like an estimate. If it is a nice, public relations-oriented company, they might do it for nothing if you promise to let them move you when the time comes. Possibly, a small amount may be charged. Pick a city and tell the moving company that you are thinking of moving there and would like their estimate. They will give you a written estimate which will include the approximate volume, weight and distance. Presto, you have the information (volume and weight), which you can then use to compute the cost of moving your household goods anywhere in the world with a call to any moving company. So, armed with this information, you can then take the cost factor into consideration if you received an offer in Chicago, or any other city in the country.

The military leader candidate has a distinct advantage here over the non-military candidate. Household goods will generally be shipped at government expense when you separate from the service. If you are in the military, you should be aware of your authorized limits. Military regulations change so often that it would also be wise to take a trip to the transportation office and find out from an informed person what your limits and rights are. Don't rely upon hearsay or word of mouth. It could be very expensive in the long run.

As a point in passing, let me caution you about using your "free" military move prematurely. If you are separating from the military in San Francisco and want to return to the East Coast to look for a position, have your goods put in storage on the West Coast to be delivered to your new address when you have found employment. If you move them back to your parents' home, for example, and unpack them, you may find you will have to pay to ship them (if the hiring company won't) to wherever you end up. It is inconvenient, but generally not impossible, to live out of a couple of suitcases for a few weeks. If that **is** impossible, then do not unpack all the goods, as a major part of the moving expense is the packing and unpacking of boxes. You should note here that some companies do not realize that you have a free relocation upon separation from the military. Make it your business to be sure they know. It gives you a competitive advantage.

While the question of who is going to pick up the costs of moving your possessions is an important one to you, it is generally a matter of **secondary** importance, both to you and the recruiter. You should not bring it up as a question in the initial interview– the information should be included in the details of an offer and you should get the facts at that point, if they are not cleared up earlier.

However, if you are able to move in a couple of suitcases, make that fact known– in your cover letter, on your resume or in the interview. It is possible that this could make the difference between getting or not getting an offer, particularly if you are in competition with local applicants.

The whole idea is to make yourself as attractive as possible to a prospective employer. While some companies do not think twice about the expense of relocation, you cannot, at this point, assume that will always be the case.

If you interview with a corporation or organization of medium to large size and scope, you may meet with a person called "the recruiter" as your initial contact with the company. It is important to understand this individual's relationship to the total organization as well as the methods used to accomplish objectives.

Depending upon the organization, a recruiter may have responsibilities other than those of interviewing. Some recruiters are members of general personnel sections. In other organizations, a recruiter evaluates applicants on a full-time basis, and plans to meet future staffing requirements.

For the purposes of continuity, I am going to classify these two kinds of individuals as "recruiters," as opposed to "interviewers." A **recruiter** has recruitment as a **primary** function; an **interviewer** recruits as a **secondary** function and is mainly a sales manager, engineer, personnel manager, etc. All recruiters are interviewers; all interviewers are not recruiters.

As an outsider, you might think that a recruiter would be assigned to this position after close scrutiny of educational qualifications and business experience, with priority going to degrees in psychology or personnel management and years of experience with the organization. That is often not the case. The position of recruiter is often filled by a top-quality person from within the organization who has sales experience or capability. A recruiter who has been in sales for several years has developed a strong understanding of the organization, including how the product/service is marketed, the philosophy of the employer, the real qualities necessary to be successful there, as well as a good understanding of the competitive organizations within the industry. In short, this sales/recruiter person has been exposed to many levels of the organization and has this experience to draw upon when evaluating prospective employees.

An interviewer who is part of a personnel department or has other duties and has not been tied as closely to as many levels of the organization, could also be a very searching and thorough evaluator, but simply bases that on a different perspective. When you meet this individual, whether a full-time recruiter or a part-time interviewer, you should understand that perspective.

After your initial contact, you will most likely see very little of the recruiter, whether you are hired as an employee or not. This should tell you that there is a large degree of detachment from the recruiter's evaluation of you. The decision to have you talk with the next person in the chain is an objective one, as the recruiter compares you with the many other applicants faced daily across the desk.

Your goal is to put some subjectivity into the interview decision, to create some excitement, and make the recruiter feel personally involved in you as an applicant. To help generate this excitement, you will use several different methods explained in detail in other sections of this Handbook. Before you use those techniques, you should know more about the recruiter's basic motivations and take a closer look at his or her related tasks, so you can coordinate your efforts and not step on any toes in the process. Here are a few points of preliminary interest:

First, you should understand that recruiters have strong feelings about the organizations they work for. Most good employees display loyalty and commitment to their employers, but most recruiters have even stronger feelings than the average. They must. Their job description calls for responsibility for recruitment of employees in a large geographical area, and, consequently, they often travel much of the time. Being "on the road" is perhaps initially a lot of fun– staying in nice hotels, eating in nice restaurants, and meeting many new and interesting people. But it all gets very old very fast. The hotels become impersonal, all the food tastes the same, and the constant change becomes routine. A person in this position is motivated to continue for many reasons, but one of the most important must be strong positive feelings about the employer and willingness to sacrifice home life for the opportunity to do the job. If this sounds similar to the story of the traveling sales rep, it is, but without one important factor: recruiters are seldom paid a commission or bonus each time they find an employee for their company. They are generally paid a straight salary and do not have the financial incentive that drives most sales reps who travel extensively. They are motivated by their desire to do a good job for a company they hold in very high regard. Remember that fact whenever you are interviewing with a company and want to suggest to the recruiter that there are some "deficiencies" in the company represented.

The recruiter has another motivation besides respect for the company: to be promoted. Somewhere, someone is keeping score of results. The total number of applicants interviewed and approved, the number eventually given offers, the number of offers accepted, the number of those who have done well with the company (as well as the number of turkeys), and the number who have quit or been fired are a direct reflection on recruiting ability. And recruiters are under constant pressure to find the best possible applicants. If it goes well, promotion into higher levels of responsibility is in the future; if not, who knows?

A recruiter does not hire you. The recruiter recommends that you continue on to the next level for more interviews, and if the organization is large enough, might never see you again. Even when the recruiter has found a good applicant, the applicant is not guaranteed an offer. The more exacting needs of the actual department or specific job must be considered. Since the recruiter is often looked at as the resident expert on people evaluation, that stamp of approval can be valuable to you. A recruiter is a screener in the sense that well-known general requirements are used as a basis for making the match, and you will then be referred on to someone who has the ability to fine-tune the evaluation to meet the specific requirements of a particular position.

The pressure to find good applicants coupled with the travel and repetitive nature of the work makes the recruiter's job a monotonous one after the novelty has worn off. But the recruiter cannot show this boredom, ever. If the primary duty of a recruiter is to find good applicants, the secondary one is that of public relations specialist.

Each person interviewed is a potential employee (at best) or a potential customer and advocate of the company if the interview does not result in employment. Everyone interviewed cannot be hired, and if the ones not hired are mishandled, an unbelievable amount of damage can be done to the company. Let's use an example: a recruiter talks with six candidates a day, not an unreasonable number. If the averages allow one of those six to be referred on to the next step, that means, on a five-day interview week, there are 25 rejected applicants and a couple of candidates who were referred on but were rejected later. Let's estimate that there are 27 unhappy applicants generated per week, a fairly conservative number. Let's also say that this recruiter interviews 50 weeks out of the year. That equals 1,350 unhappy people per year, per recruiter. And if there are ten recruiters in this fairly large organization, we are talking about 13,500 disgruntled people per year. If these people are sufficiently alienated, they will not buy the company's products and they are going to tell their wives, fathers, husbands, sisters, aunts, uncles, and grandparents how poorly they were treated. Before you know it, there is a huge drop in sales and the company is announcing reorganization plans to cut staff, all because the recruiters were not working with their secondary purpose of public relations in mind.

That example is perhaps a bit farfetched. I doubt there is a major organization with ten dumb recruiters ruining sales. I used the example to draw your attention to the fact that recruiters cannot allow their personal feelings of boredom or unhappiness to influence the manner in which they work with applicants, and to an even greater extent, the manner in which they handle applicants they must reject. (There is more on this subject in the section titled "The Employer's Sale.")

At one point you will run across a recruiter who cares little for the public image projected by the organization, and who treats applicants with less than common courtesy. When you are faced with an obnoxiously self-important recruiter, don't damn the entire organization. Do your best in your interview and use some common sense. While you are reading section 26, "The Negative Interview," keep this recruiter in mind. You can predict a tendency to use the negative interview but not necessarily for the valid interviewing purpose you will read about in that section.

If you feel you have been grossly mishandled, and you know you are not going to be allowed to interview any further, write a letter to the company president. Write it in a civil tone, explain the facts and your case. Do not ask for another chance to interview. That will sound as if you are blaming the recruiter for your failure. If the company is PR-minded, it will investigate and possibly ask you to interview again, with a different recruiter, one hopes. At any rate, the management of that company should appreciate hearing about these goings-on and do something about them.

Most recruiters are well-adjusted and gregarious people. They are easy to talk to and present a good image of the company they represent. You will notice that I have discussed the recruiter in this section and have offered little about the interviewer (that person who is a manager in the company who must interview you later). We will discuss that person in later sections, for the interviewer is equally important and makes up a large number of contact points you will have on the road to employment offers.

I hope you are more familiar with the recruiter at this point. This is a person who has strong attachments to the organization, and is motivated to travel extensively and pay the price because of those strong feelings and the desire to be promoted. During your first interview, the recruiter's task is to evaluate your fit to some general criteria. After evaluating your assets in relation to the job criteria and against other candidates' attributes, the recruiter decides, **very objectively**, whether or not to refer you on to the next step. A good recruiter is very easy to talk to, and very much aware of the secondary role as a link between the company and the public. For the uninitiated candidate, it is often difficult to tell whether an interview has gone well or poorly because of this PR side of the conversation.

Perhaps you are beginning to recognize that an interview is conducted in a selling environment. You are selling yourself to someone who may, or may not, be interested in the product. You need every inch of ground you can get and to gain it you must exploit a basic and successful principle of sales: use the negative (didn't, couldn't, wouldn't, etc.) as seldomly as possible. In short, stay positive in your syntax.

Q: "When did you decide upon Business Administration as your major?"

A: "I didn't choose it until my junior year."

Reaction: "What took you so long?"

Look at a different answer to the same question:

A: "I entered college with a wide range of interests, and chose Business Administration as my major in my junior year, because I felt it would give me the broadest base from which to build a career."

R: "That was a very complete answer. Let's go on to the next question."

The second answer takes longer, but you can cover more ground on **your own terms**. A recruiter is trained (if by nothing more than daily practice) to look for holes in answers. At the same time, a recruiter will react or respond as much to the **words** you use as to the content of your sentences. A very natural reaction to the negative is to question it. Analyze your own thought patterns to the statements below. If someone said to you:

(you'd respond)

I didn't like it...why not?

I disliked school....................................what did you dislike?

I can't go out with you .... why not?

I don't like military types..screw you!

It is practically impossible to drop all the negatives from your vocabulary, but you should make a conscious effort to express your thoughts positively whenever possible and avoid this questioning reaction. Another example:

Q: "Why are you leaving the military?"

A #1:   "I didn't like the structure, and I didn't get recognition for how hard I worked."

R: "What didn't you like about the structure (and how is it really different from ours), and what do you mean by getting recognition?"

A #2:   "The military structure is based on accrued time in service and in rank. Unless you are extremely bad, you will be promoted along with the next person of your rank. I want more than that. I want to be promoted and recognized for my individual productivity."

By using a positive and assertive approach, you control, to a great extent, the direction you allow the questions to take. By being positive, you eliminate unnecessary questioning of certain facts, as demonstrated in the previous example.

There are two sections in this Handbook which I consider essential. If you can master the concepts offered in these two sections alone, you will be way out in front of the competition in your interviewing. This is one of those sections, so wake up, get another cup of coffee, close out all distractions and pay particular attention.

To start off easily: RULE #1 is: never, ever, answer a question with a simple "yes" or "no." Obvious? Certainly. But what is not so evident is that many people untested in the interview situation are so tense, so keyed up, and so expectant of the terrible things to come, that they will do this:

Q: "Are you just getting out of the military?"

A: "Yes." (God, **when** will we get to something important!)

They do not stop to think that if the interviewer really wanted to know whether the applicant was just getting out of the military or not, the answer is probably on the individual's resume.

Many questions in an interview, and particularly in the first few minutes of the meeting, are questions with obvious answers, answers which could readily be taken from the resume or application form. They are asked to put you at ease – a time-tested technique used to get you relaxed and the interview started. And most applicants will do **exactly** that; relax as they answer the simple yes/no kinds of questions.

Q: "Where did you go to college?"

A: "The University of Kansas."

Let's discuss another concurrent phenomenon. A recruiter talks to lots of people. That is fundamental to what we do. If there are 20 interviews on one short recruiting trip, I probably return to my office and quickly review the files the next day, before the notes get cold (while **you** shouldn't take notes during an interview, recruiters generally need to). In making this review evaluation, two extremes naturally stand out in my mind: the very good applicants and the very bad applicants. Those two areas comprise about 20% each of the total, which means that 60% fall into what I call the "gray zone", those applicants who really are not **bad**, but just don't get anyone excited. If the recruiter is really hard-pressed to fill positions, that gray zone will be used to find the additional people needed. But that doesn't happen very often.

Obviously, you want to be in that upper 20%; the people who create excitement and stand out after listening to all of those people answer questions.

Now, just think for a second. Are **you** really impressed by people who go around answering questions with "yes" or "no" and give curt, superficial answers? Unless they are well-known rock- or filmstars, I'd guess not. And for the same reason recruiters are unimpressed: because such people are overshadowed in your experience by articulate people who tell you about themselves in more detail, with a sense of humor and perspective. Those are the people and the personalities which stand out in your mind and are remembered after the initial exposure. So the #1 RULE is: no "yes" or "no" answers, ever.

If you can't answer a question such as "Are you just getting out of the military?" with a "yes" or "no", how should you answer it?

Glad you asked, because it brings us to the second point: RULE #2. A recruiter is not interested in what you have done, but rather why you have done it. Talk about the **why's**, not the **what's**. That sentence is so important, you should read it again, underline it, and make sure you understand what it means.

I'm going to show you how application of this rule can bring you to the top 20% and also help you to **control** the content of the interview to your advantage. Let's pick a question to use as an example.

Q: "When did you pick your major in college?"

98

Well, you know not to answer the question with a yes/no response, so:

A: "In my junior year."

is out.

You should immediately recognize the question is not "**When** did you pick your major?", but is really "**Why** did you pick your major?", and if you recognize it as such, your answer might come out something like this:

A: "I really started getting interested in accounting in my sophomore year, and I had the opportunity to work with an accounting firm as an office helper during the summer vacation. When I came back to school in my junior year, I declared my major as accounting. It was a good choice for me. I enjoy the detail and discipline of the science and hope to build my career on an accounting base."

Let me assure you that an answer like that is going to make your interviewer sit up and take notice. After all, "my junior year," "my senior year" and "my mother told me to be an accountant when I was in high school" are answers to that question that are expected from almost all applicants interviewed.

That's fine, but if your major was chosen late in your junior year, and only after you had declared your major as (in order) history, English, sociology and accounting, you have a very different situation and need to approach the entire answer somewhat differently. When you read the section on "Direction," you will know that having four majors in three years is not very positive, and telling the interviewer is going to do more harm than good. But yet, you cannot lie.

What to do? Explain **why** you declared your major in accounting in your junior year **on your own terms**.

Q: "When did you choose your major?"

A: "When I entered college, I had a wide variety of interests. I experimented and took many different courses and chose accounting as my major in my junior year, because I felt it fit my personality and goals better than any other field I investigated."

This concept is also closely related to the material discussed in the section on negative sentence structure. Both techniques establish control over the information allowed to come out in the interview.

Let's look at some other examples of questions using the principle of "why, not what."

Q: "How did you get your military commission?"

A: "Back when I was starting college, ROTC was mandatory for the first two years at my University. I had the opportunity to drop it after the initial commitment, but decided that I would continue when I was offered a two-year complete scholarship by the Army ROTC program. I graduated and was commissioned as a reserve second lieutenant." (You could go on to explain why you picked Army ROTC.)

Q: "Where are you from?"

A: "I was born in Boston and lived there until I was in the seventh grade when my father retired from his job as Production Manager for Acme Industries. We moved to Presque Isle, Maine, and I lived there until I completed high school and left for Georgia Tech." (You could go on to explain how you benefited from living in a rural and large city environment.)

And:

Q: "How is the interviewing coming along?"

A: "Very well. This is a very interesting and exciting process for me. I've had the opportunity to speak with representatives from several excellent organizations and look at their (sales / engineering / production) opportunities. I have really been delaying further steps with them, though, because I have been very interested in your organization for a long time. I want to see if there is a match here before pursuing any of the other options open to me."

You do not want to take this rule to its extreme and sit there talking like there was no tomorrow on a relatively unimportant subject. But I'm sure you will agree, these examples and the answers outlined add much more to the interviewer's knowledge of you than yes/no responses to the same questions. Additionally, mixed with a little foresight, you can tell the interviewer what **you** want known by expanding the answer in a positive direction. This method of answering the recruiter's questions with depth has the extra benefit of attracting the interviewer's attention from the very beginning of the interview, where often superficial questions are asked in order to put applicants at their ease.

Try it. It works.

There is no way that reading this Handbook will give you good judgement in your life or in your interviewing if you don't have it at this point.

I define judgement as the process of evaluating a situation, determining the variables and possibilities and then taking the most appropriate action. This is necessarily a subjective process where you need to appreciate and balance the various possibilities and outcomes. There are some things that are either there or not there. All the reading you can do will not give you good judgement. I want to cover only one point, so you might recognize a common shortcoming and do something about it if you think you might make the same error.

Q:	"Why did you join the Air Force?"
A:	"To stay out of the Army and avoid being drafted into the Infantry with all those grunts."
R:	"Why are you insulting me, a former Infantry officer?"

In this example, poor judgement could lead to a very uncomfortable situation or, more likely, immediate dislike, which could be extremely difficult to turn around. Of course, there is nothing to be gained by being a wishy-washy, unconvincing applicant giving broad, general and vague answers either. So what is the best approach?

Two little words: **FOR ME.**

A:	"For me, the Air Force was the best choice. I was able to choose from among six specialties and three locations."

I think it is an honest answer to the question. What made the Air Force attractive to you might make it unattractive to someone else.

Before you express an opinion about something, pause to consider the consequences which may result if the interviewer (who would probably never let on) had reason to have a very different opinion. The example above is fairly obvious, but the real situation could be more subtle:

Q: "How important do you think a college degree is in getting ahead in business?" or

Q: "What do you think about our competitor's product?*

Where your opinion is requested you set a neutral stage when you use those two little words. You don't **need** to get into an argument, spoken or unspoken, with your prospective employer.

---

* Have I really got you thinking on this one? The truth is, most managers have a very healthy regard for their competition's product. In fact, in many industries, it is very safe to assume the manager worked for the competition at one point. Your enthusiastic annihilation of the competitor could backfire.

In the section "Depth," I mentioned that there were two sections in this Handbook which I considered essential. "Depth" was one, this is the other. Frankly, this section could be subtitled, "Tell them what they want to hear," if you view the contents superficially. Take this section into consideration with the total contents of this Handbook and in its proper perspective. To simply tell someone what they want to hear implies dishonesty or lack of conviction. You are not interested in presenting either of those personality traits in an interview.

In this section, we will discuss the principle of empathy. Specifically, empathy with the interviewer, and how that empathy will lead you to better and more successful answers. Before we start, you should flip back to the section on self-analysis. Review the list you made of your positive and negative characteristics.

By way of example, I am going to pose a question to you. You have seen it before, and answered it in a different situation:

Q: "What is your most positive personality characteristic?"

If you answered in the following manner, using depth:

A: "I feel my most positive quality is my ability to set a goal for myself and meet that goal through hard work. I have used that ability continuously in my military career. I was given general goals by my superiors and carried out those missions using my own standards and a lot of extra hours to get excellent results,"

you should pat yourself on the back, because you have given a good, solid answer, and better than most. But, under some circumstances, it might not be good enough. Here is my point:

If the interviewer is looking for a good production manager, there exists (possibly right there on paper on the desk) a list of attributes sought in production manager applicants. Possibly, this list has been developed formally by an industrial psychologist. Just as possibly, it was developed by the recruiter simply remembering the characteristics of successful production managers hired in the past.

Here's the tricky part: **you need to use that list in the formation of your answers.**

Obviously, you cannot get inside that interviewer's head and read that list, so you must do the next best thing and create your own list. Before every interview, you should take a few minutes and jot down the qualities which typify the position for which you are applying. You will use those qualities in your answers in this particular interview.

As an example, off the top of my head, I'd list the following qualities as important requirements for a job in production management: ability to work well under the pressure of a production quota, the ability to make decisions well and quickly, the ability to relate well to people, the ability to work well in a semi-structured situation (i. e., a situation that changes little physically and is somewhat routine from day to day, but can erupt in unexpected challenges), the ability to understand the mechanical side of machinery operations, and a familiarity with or knowledge of fundamental (union) contract principles. Technical knowledge may or may not be a requirement, depending on the level of the position.

And, if I were being interviewed for a position as production manager, my answer to the question above would go something like this:

Q: "What is your most positive personality characteristic?"

A: "I feel my most positive quality is working well with people in a high-pressure situation. There are several examples I could use in my military experience, but the best example is when my unit was given 24 hours notice that we would perform a live fire exercise for a visiting Congressman. I was able to organize my troops, go through several practice runs, changing the program three times to meet the situation, and ended up giving an excellent demonstration delivered on time. Other units would have needed a lot more time, but even with the pressure, I was able to get the job done with teamwork and good humor. "

I have hit what I consider the most important characteristic the interviewer is looking for: the ability to work in a pressure job situation and get the job done well. I am projecting my answer onto the needs of the interviewer.

You might object to this method on the grounds that the question was, "What is your **most** positive quality?" and, by definition, that means there can be only one best quality and you must name it. But let's go back to the list of positive qualities you prepared earlier. Look at the first quality you listed. Now look at the fifth. Is there that much difference in the intensity with which you feel the first and the fifth fall into that order? I would think you could list at least five personal qualities you consider strong enough to interchange with that "top" quality.

Let's look at something else. You have just interviewed, successfully, of course, with a person for a position in production management. You have given the answer above. You walk out of the office and up the street where you are scheduled to interview for a position in sales leading to management. As you wait, you jot down the five or so qualities you believe typify this sales position: ability to relate well to a wide range of people, ability to work in an unsupervised and unstructured environment, basic intelligence, good image, high degree of organization, and self-motivation. You go into the interview and are asked:

Q: "What do you consider to be your outstanding personal quality?"

A: "I feel that my most valuable quality is my ability to work in a situation without supervision. My military commanders felt that, too, and frequently gave me important projects that needed to be done but were not in the general job description of a supply officer. A good example of what I mean would be when I was assigned the task of setting up a new officer's club on base. I went through all the regulations and, without supervision, set up a club in an old building. The only time the base commander was involved was to select the furnishings. I really enjoyed doing that project because it allowed me to take my own initiative and get it accomplished. "

You identified the key trait here as the ability to get the job done without close management by your supervisors, a trait that is considered valuable in an outside sales situation.

If you have done the proper research, it will not be difficult to jot out these five typifying qualities. Remember, they will change from position to position. Although you will find some common to many jobs (e. g., ability to relate to people), you should highlight the **unique** quality rather than the common.

Let's take this principle one more step. Think of a list of your hobbies so you can answer this next question. I'll set up a list of some theoretical ones: you like to ski, you like to work on your sports car, you like to play racquetball, work in the garden, read interesting books, develop your own photographs and build cabinetwork in your house. This is a fairly diverse list of avocations, but you feel prepared to offer any one of the hobbies listed as one you enjoy very much. You go to an interview for production management in a heavy manufacturing environment:

Q: "What do you do in your spare time?"

A: "I really have a wide variety of interests. I'd say that one of my particular favorites is working on my sports car. I have always had a mechanical ability, and I enjoy doing some fairly sophisticated work such as replacing the clutch or relining the brakes. "

The hobby chosen from the list is one identified as having some applicability to the job applied for. Walking out of that situation and into the sales leading to management position, asked the same question, your answer is going to be:

A: "I enjoy playing racquetball in my spare time. I enjoy the one-on-one situation, where I compete against the other person. I also enjoy it because it is a sport I can improve with a little self-discipline and application of certain principles. Aside from good exercise, I enjoy the social part of the game, too. I have a good circle of friends who play with me. "

You have identified your positive qualities of being competitive, having an ability to be self-disciplined and learn from the application of basic principles, and a social side as well. All these are qualities the recruiter looks for while interviewing for sales personnel.

Looking back, it doesn't make much sense to answer the question about what you do in your spare time with the answer, "I like to work on photography," in both the sales and production management interviews, even though photography might be your current main interest. You must name an interest that is important to the interviewer, one that possibly relates to the job at hand, but you must also answer the question with DEPTH. Look at this answer:

A: "I enjoy working around my house as a hobby. I remodeled the kitchen in my spare time last summer. It was really a big job, and I saved a lot of money. "

That might be a good hobby to pick if the job needed a resourceful and handy applicant to do odd jobs around the plant, but compare it to this answer:

A: "I enjoy working on my house. I've done everything from remodeling the kitchen to installing a new sprinkler system in the garden. Regardless of the size of the project, I make the best plans I can, buy the best materials I can afford, and really throw myself into the project. I might come home on a Friday evening and not stop working except to sleep and eat until Sunday evening. If I don't like a particular section, I will pull it out and redo the part until I am satisfied with the results. When I am finished with that project, be it a remodeled kitchen or a new picture frame on an old painting, I get a feeling of real satisfaction knowing I have done the job well and that it looks good. "

The applicant has told the recruiter that satisfaction doesn't exist unless the job has been done well, incorporating planning, managing and dedication. There are few employers who would not find those qualities attractive and valuable. DEPTH goes with EMPATHY, hand-in-hand. Using one without the other is going only half way.

Because I feel this is a very important section, I am going to close with a theoretical situation that might help you **remember** the contents more easily by illustrating the point.

Let's say that you were a multi-talented athlete, and you were about to try out for several teams. You get a telephone call from the manager of last year's World Series champion. He wants you to interview for the catcher's job, but needs to know first what you consider to be your strongest sports ability. Would you tell him how well you can throw touchdown passes? No way! He can't relate your ability to throw a football to what the team needs – someone who can catch a baseball really well.

The situation may seem comical, but that's exactly what most applicants out there are doing in interviews when they **could** choose to discuss an appropriate and applicable skill or interest.

When you are asked about your positive qualities, you will have little problem finding examples. It is an ego trip of sorts. You are thinking of all the good things you are and finding one you think is applicable to talk about in the particular situation is not too difficult. It is no coincidence that many interviewers ask first for the positive qualities, **then** for the negative qualities. Let's look at a thought pattern:

Q: "What is your most positive quality?"

A: (Thinking of being reliable, smart, good-looking and other ego-involved qualities, you give a positive answer.)

Q: "Fine. Now tell me what about you needs to improve most."

A: (Still inwardly involved, thinking of your negatives:) "I have difficulty making decisions."

It is not easy to talk about your negatives without damaging your case and giving the interviewer reasons not to hire you. You are looking inside yourself and being truthful to the point of being dumb.

My advice is to be smart. Instead of looking inside yourself, look outside and tell the recruiter something that is less damaging than the dumb answer above. A simple example of this concept would be an answer that plays upon the obvious.

A: I have been in the military for five years, ever since I graduated from college. I'm applying to you as a mechanical engineer, and I have a degree in that discipline, but I have had little opportunity to use the knowledge I gained as an engineer in the service. I hope you realize that I am rusty, and **that** might be a negative. I have spent more time with my engineering books and I have tried to keep up with what's been going on in engineering by reading *Engineering World*. Even so, I feel my experience in working as a leader in the military has developed many abilities I could never have learned from books."

You will note that the negative, which is fairly obvious to even the casual interviewer, is brought into the conversation, explained, and then turned around to be as positive as possible. Where this is possible to do, do it. It will take some time to find all these negatives and prepare answers, but if you don't take the time, recruiters will find them for you. Look at your resume or application form. Is your degree perhaps not exactly in line for what you are applying? Your experience?

Everyone has one question that they dread answering but know will be asked in an interview. Look at it as an advantage. You go into the situation knowing one of the questions you will be asked. Prepare for it! Now is the time to prepare for those difficult questions, before the interview, not as you sit there in front of the recruiter trying to think of a good reason you majored in art history.

Beware of trying to answer the question in this manner:

A: "I have no negatives. There is nothing wrong with me."

The recruiter may just ask you to walk across the nearest body of water. Of course you have shortcomings, everybody does. It's just that you are too smart to talk about the dumb ones that will get you kicked out of the interview. Cripes, you have been in the military for several years, saluting uniforms and issuing (or being issued) orders. That is not the normal way of doing business out here where people will tell you to stuff it if they don't like your approach. If you need some help discovering some shortcomings, read the section "Your Greatest Challenge." Most **civilian** managers could find many personal shortcomings in that section if they were honest with themselves.

Go back to the beginning of this Handbook and read the short list of negatives you listed there. Decide now which of those are allowable in an interview and which are *personal* negative qualities that should never be allowed to surface in an interview session. Don't ever confuse an interview with a confession.

As you sit across from that interviewer answering questions, an analysis of you and your past experiences is taking place. This is one of the most important questions being asked: is this applicant here today as part of a logical pattern or simply by accident?

In this evaluation of you in the interview, the recruiter will examine your answers to various questions to determine if your life has been directed towards the objective of working in the field and specific kind of organization you are interviewing with. In talking about the goals you set for yourself in high school, in college, in the military and in your personal development you will direct your answers toward satisfying this question.

To show clear direction is not difficult for the person who has known, since a very early age, that a particular occupation was his or her vocation. If influenced by factors which gave direction toward advertising, a high school or college career could have molded you in that direction with summertime employment and perhaps some extra-curricular involvement in the creative arts field. It is easy to present those facts and show clear direction. Military leadership is easily related to management for career continuity.

However, most of us have had many different career goals in mind during the period covered by high school, college and the military service. The average college student changes majors several times and has several different careers in mind before graduating. The attempt to show good direction becomes practically impossible if each of these incomplete goals is discussed in the interview. The solution to the problem is somewhat complex, but I intend to discuss it here, at length, because it is singularly important.

As an initial step in the process, go back to the beginning of your Handbook and examine the list of goals you set and achieved which you outlined as part of your self-analysis. In examining the list, can you find some direction? Possibly you can revise this list to show step-by-step progress towards your position today: educated, with leadership under your belt, and perhaps with some experience in the particular field you seek as a profession.

Example: your initial involvement in graphic arts, an interest developed in business in high school (in the form of participation in a Junior Achievement-type organization or other business-oriented club) paralleled by the selection of an art major for a couple of years, then the change to a business-oriented curriculum and a concentration in advertising courses in college.

This "discovery" of direction in your past should be as complete as possible. There may be some direction you will not recognize until you analyze your achievements thoroughly. It is also likely you might have to change your thoughts on some of the goals you thought were important and emphasize others which fit into the pattern. You might even re-examine the real motivations behind setting the goal in the first place.

Let's say that you have written that you went to college because everyone you knew went there, or your parents wanted you to go. These might be true statements, but looking back, with the 20-20 hindsight we all seem to possess, can you also add that you went to college because the degree was/is a general requirement for the career level you wanted to establish, or the standard of living you hoped to achieve. We are starting to see gray areas from different perspectives. What you believed to be true at the time changes in the light of the overall picture at present.

At this point, we should talk about the subject of honesty in an interview. At no point am I advocating that you lie about yourself or your qualifications. That is out. You can be fired, or face legal action, if an organization hires you and finds out later that you represented yourself falsely to them when you were interviewing. Most large organizations have a statement to that effect on the application form which you must sign at one point. You must remain truthful in your presentation of the facts.

116

The question (as I see it) is: where does being honest stop and being stupid begin? To what extent should an applicant tell the recruiter facts which are honest, but damaging (particularly when these facts are personal and open to wide interpretation)?

To use a parallel situation, if you wanted to sell your sports car, you'd run an advertisement and eventually receive calls from people who want to inspect the vehicle. Would you tell those prospective buyers the minute you opened the garage door that there was rust under the right fender? No way, not if you were a good salesperson. The method would be to point out the positive aspects, leaving the negative to be discussed later, when the buyer had a stronger desire to buy the car and would not react as strongly to what could be seen as superficial, negative facts. You note that I used the example of rust and not that the transmission was missing, right?

You aren't a used sports car, with or without rust, but the principles of the sale are the same. You do not **offer** negatives if they are damaging to your presentation. If you feel you must mention these negatives for moral reasons, make sure you know how the facts will affect the interviewing process and deliver them when you know the interviewer is somewhat impressed with your talents and will pay less attention to your negatives. Normally, that is not going to happen in the first interview.

Now let's look at a different situation. You are selling the sports car again and this time the buyer asks, before you even have the chance to open the garage door, "Is there any rust on this car?" Your answer must be truthful, but with a positive emphasis. "Yes, there is, but it is minor and could be repaired easily." You are doing the same thing here that we did in making a presentation of your negatives; trying to turn them around to be as positive as possible. Now, let's take these thoughts and return to the subject of direction. Let's use an example, a rather obvious one, which should be answered with direction in mind.

If you were asked the question:

Q: "Why did you go to college?"

117

The honest answer might be:

A #1:   "I went to college because my parents always directed me toward higher education and all my friends in high school were planning to go to college, so I went, too."

That is an honest answer, but it is dumb. It is damaging for reasons I do not need to explain to you if you have been reading this Handbook carefully. With a certain amount of effort and, frankly, rearranging of your thoughts, your answer might come out in this form:

A #2:   "I grew up in an atmosphere of higher education. Both my parents went to college and I made the decision to go when I was in high school. I knew that if I were to maintain my familiar standard of living and realize the goals I had started to set for myself, I would have to meet the basic requirement of a college degree, etc."

This example could and probably should be expanded to explain what the goals were and why you set them, but we'll talk about that later in this section. The difference between the two answers is evident. There is no question which of the applicants went to college as part of a design, and which went by accident. Is the second example an honest answer? That depends upon the particular case, of course, but in general I'd feel safe saying it was honest, salted with a good sprinkling of smarts. Can you live with that?

Let's look more closely at the methods of demonstrating good direction. The best way, perhaps, is to ask a question often used to find direction, and examine a couple of possible answers.

Q: "Tell me about yourself." (That is not really a question, is it? It seems more like a challenge.)

If you remember the principles of depth and empathy, you will first recognize that the question is really not, "Tell me about yourself," but rather, "Sell yourself to me," or "Why should I hire you?" And, secondly, you will not want to answer the question in this manner:

A: "I was born in Berkeley, raised in Berkeley, went to Berkeley High School, went to the University of California at Berkeley, spent four years in the Army, and here I am."

Again, I used an exaggerated example to demonstrate the content of many uninformed applicants' answers to the question. You know much better than to answer a question in that superficial manner; you'd never let that happen. But here, in trying to show clear direction, the situation calls for more than just giving good depth in the answer. It is in this kind of direction question that you might be dumb and include too much detail, some of which could be harmful, in your attempt to answer the question with depth. **Do not confuse depth with extreme detail!**

A: "I was born and raised in Berkeley. I have three brothers and a sister. The sister is older than me and her name is Jane. My youngest brother is 15, and his name is Jim. My other brothers are named Jack and Jeff. Jack is 17 and Jeff is 19. We live in a three-story house, etc. I finally decided to go to the University of California at Berkeley and started there on September 17th. I first decided to major in history, and then decided that I didn't like that because there was too much reading. I decided to major in business...."

Boring - and possibly very harmful. You have revealed (in this case) your wavering conviction in deciding upon a major. That could lead to some rather pointed and truly negative questions. Answering the question, "Tell me about yourself" is an exercise in interviewing techniques. You will need to use all the empathy, depth and direction you can muster in presenting a complete answer. I'm going to give an example of how it might be answered, and you will notice that the answer is extremely long in relation to the other answers I have used as examples.

The length of the answer should be determined by your own estimation of the situation. If you are sitting down to interview and the recruiter says, "I have been talking all day, and I'm tired. You don't mind if you do some talking, do you? Why not just take some time and tell me about yourself." Your answer might be as long as fifteen minutes, maybe even longer. If you have been on earth for over twenty years, surely you can talk about yourself for fifteen minutes. If, on the other hand, you are faced with a short interview, and note a hurried expression such as, "We don't have much time. Instead of me asking you lots of questions, why don't you tell me about yourself?", you will want to keep the answer shorter, but still with good depth. Do not sit there and read your resume or application form to the recruiter.

Let's look at a particularly good example of an answer to the question, "Tell me about yourself."

A: "In order to give a good and complete answer to that question, I'd have to go back to the point in my life where I first really took on responsibility. I was born and raised in Berkeley, and came from a large family. Even though my father had a good position as an engineer with XYZ Company, when you start spreading that income over seven people, there are some things that are simply not available. I wanted the extras that my parents couldn't afford, so I got a job as a paperboy for the local newspaper. As a matter of fact, I ended up with the largest newspaper route in the city and made some good money while I was going to junior high and high school. I mention this because *(great control!)*, as a paper carrier, I really was exposed to a business atmosphere for the first time. It was a good responsibility for a teen-ager, and I got a lot out of it. As I was getting started in this job, I was also involved in Scouting, and worked toward becoming an Eagle Scout. My parents were a big influence in my reaching that goal because I had developed a strong interest in athletics as I moved from junior high to high school. I managed to do all three things at the same time: participate in the leadership of Scouting, compete in athletics and run a paper business. As I entered high school, I understood the pros and cons of going to college. I decided early that I wanted to go, and to go to a good one. I dropped the Scouting and the paper route, although I was still heavily involved in sports. I knew that I would need to get good grades in order to have options in selecting my college, and I worked hard to get them. I also became involved in other extra-curricular activities and found I enjoyed the political offices to which I was elected. In my senior year, I applied to six colleges and received acceptances from five of them. I chose XYZ University because I felt it was a good school and, since finances were still a consideration, I could live at home while attending. I experimented with several different kinds of courses and chose business administration as my major in my junior year. I felt it would give me a broad base from which to build a career and felt then, as I do now, that there is a lot of opportunity for me in the general field of business. While in college, I worked during the school year and during summers for a local manufacturing business. I was exposed to many of the day-to-day operations, and that exposure has excited me about the possibilities of applying the principles which have made me a success as a student. I feel I can also be a success in business."

I could go on, but think I have made the point. If you review that answer, you will find that you can draw a line through just about everything included there and point it towards the objective of a business career. It is showing excellent direction.

Just as important as what was included is what was left out. I did not include the fact that the applicant majored in history, then English before choosing business administration. I did not include the fact that the applicant went to Europe for six months before starting college. I didn't include the fact that the applicant was kicked off the football team for indecent exposure on the fifty yard line at the homecoming game. These are negatives either in direction or character, and to offer them up for judgement in the interviewing process is a poor decision and poor salesmanship.

If the applicant is asked at the end of the presentation, "I see you graduated from high school in June, but didn't start college until January. What did you do in the interim?", this applicant must be truthful, but answer in a manner which will do as little damage as possible to direction:

A: "I forgot to include that. The opportunity presented itself to go to Europe. I was at first reluctant to interrupt my studies, but decided to take the time while I had the chance. I'm glad that I did that traveling. It exposed me to many things I had only read about in books. As a matter of fact, it made me even more certain than ever about my decision to go to college, having seen how many other people live who do not have the opportunity to change their life through education."

When you are involved in your career and are down the road a few years, your direction will be judged relative to your age. If you are a thirty-year-old professional and have had three jobs in the last four years, the most lucrative of which paid you $22,000 per year, the direction you show is poor, regardless of the lines you can draw from step to step, and a good interviewer will tear you apart with pointed questions. At the entry level, the majority of your achievements and conscious goal-setting have been in high school, college and in the military. It is for that reason you must present a clear and well-directed picture of your personal progress. You have relatively little experience with which to show your pattern. Some entry-level applicants object to my request for high school activities, feeling these details are insignificant. I feel a resume which shows only leadership on the college level is lopsided and does not show the logical development of the applicant through those few years available to show direction. By the time you are six years into a career and want to change jobs, your resume would, of course, have enough material to show development of career abilities, so you would not need to show high school direction.

Work on your presentation of direction by answering the question, "Tell me about yourself". The time you invest now will help you in answering all kinds of questions later.

I would have to classify this kind of question as the hardest to detect, and therefore, the hardest to answer well. It is the misleading question you remember as you are falling asleep the night after the interview and realize you did not see the **true** question when you were asked it. Then you start playing the "if only" game: "If only I had said this instead of that." Everyone plays it. Forget the past. Let's look at preparation for the future!

Misleading questions are often used to test your judgement, and if you have not read the section in this Handbook on "Judgement," you should do so before reading this section.

Misleading questions are **generally** asked in the second half of the interview when you have developed a rapport with the recruiter. You feel the interview is going well, and your guard is down. You may be in the position of trying to impress the recruiter with your abilities, interest and personal goals even more, because you have this beautiful momentum going. Let's look at a typical misleading question:

Q: "OK, Jack, let's say we hired you into this position. You know it is entry-level and that there is plenty of opportunity to progress in our organization. Tell me, where would you see yourself in five years with us, in terms of position and income?"

Your answer, under the circumstances, might be:

A: "I'd like to be making $100,000 per year and be close to a vice presidency."

Reaction: That's impossible from what I've seen in this company, and what Jack is really telling me is that he has either very unrealistic goals, or that when he comes to work for us and sees it is impossible to move that quickly, he will probably become disillusioned and quit.

You have blown the entire question by telling the recruiter what **you think** is wanted (high goals), and you are speaking from a position of ignorance. Now really, what is the **honest** answer to this question? Can you really tell me where you are going to be in five years? I doubt it. In terms of money, the uncertainties of inflation and the economy make it almost impossible. Can you really tell me what you are going to be doing? Doesn't that depend upon how the organization reacts to your successes and demonstrated abilities? Do you **want** to be able to tell me what you are going to be doing on a programmed basis? (If you do, you might want to reconsider your military career opportunities.) Let's look at a better way to answer the question:

A: "Mr. Jones, it would be very difficult for me to pinpoint where I will be in five years. I feel there are many different considerations to be taken into account. I can say that I'd hope to have progressed in the amount of responsibility I was assigned, and that my scope of management would have increased in proportion to my demonstrated abilities. I have to believe that the money will follow."

So you have answered the question without setting a goal for yourself. More importantly, you have answered the question honestly. It will have the ring of truth, and you will not have the additional worry of getting your statements crossed up later in the interview when the recruiter tests your allegiance to that $100,000 goal.

It may be that this is not a direct enough answer for the interviewer, and a chart is pulled out of the drawer, showing the average time from position A to B to C. Your answer should still be about the same. Those are average figures. Recall the story about the man who drowned in the river that had an average depth of two feet? My question would be, in answer to the chart: if those are average figures, what is the fastest the promotion has been achieved? And **that** would be my goal.

Here is another kind of misleading question:

Q: "Are you interested in sports?"

A: (Thought: this recruiter must be interested in sports, or the question wouldn't have come up. I'd better show an interest, too.) "Wow! I love sports. Every chance I get, I'm out there at the ball park or playing tennis or on the slopes. Yes, yes, I just love sports."

R: If this airhead likes sports that much, he will probably spend all his time goofing off instead of working. I'd better pass and hire that other candidate who has different priorities.

Again, you have blown the question by attempting to tell the recruiter something you think should be the answer. And, again, all it takes is a little common sense to qualify the answer and make it more acceptable:

A: "I certainly enjoy sports, both as a participant and as an observer. I spend a lot of time perfecting my tennis game. But tennis is my avocation, and I pursue it when I have finished the job at hand. My sports interests are just that - interests."

I hope you will recognize the more subtle misleading questions:

If you had all the money you needed, what would you do for a career?

How do you feel about labor unions?

At what point in your life will you consider yourself a successful person?

What has been your most significant accomplishment?

These are all misleading questions. They are typified by being asked when a good rapport has been established and by having two answers: the emotional one you think the recruiter wants to hear, and the honest one, made with good common sense and judgement.

If there is a particularly good example of the "game" involved in interviewing, it is the negative interview. There are few interviewers able to give a totally negative interview lasting from 30 to 60 minutes, so often this "game" takes the form of a short series of questions during a normal interview situation. Even then, it takes a fairly sophisticated recruiter to carry it off well. It is most often used when the interviewer feels you are not being truthful, or that you are not totally committed to what you say. Let's look at a typical negative exchange:

Q #1: "You know, John, I am unable to understand how a person could go through college at XYZ State, major in history and come out with just a 3.0 GPA. Why did you do so poorly in school?

A #1: What do you mean, poorly! 3.0 isn't bad! As a matter of fact, I think it is pretty darned good! What was *your* grade point average in college!!!???

You have just answered the question in an argumentative and immature manner, doing just what the interviewer was trying to get you to do. You are being reactive and defensive. Let's look at the question from the interviewer's point of view and see what's being accomplished. The question is asked for one purpose: to get your reaction. The theory is that you are going to react to this kind of question in the same manner you will react to a customer, co-worker or manager who gives you a little trouble. Are you defensive and argumentative? If so, an employer will not want to put you in a position of responsibility where you must be depended upon to work effectively with people. Let's look at some examples:

Q #2 "You graduated from XYZ College. I'd never hire anyone from there. It is filled with rich kids who have a free ticket, and a check from their Daddy to pay the expenses."

Q #3   "You decided to go into the Army. Was that
                    because you couldn't get into the Air Force?"

Q #4   "You don't really want to go to work at this job,
                    do you? You have to get up at 5:00 AM and call
                    on your first customer at 6:00. You drive all day
                    from account to account, get home late, and
                    have work to do on Saturdays so the
                    paperwork gets into the office by Monday. We
                    don't pay well, either. You don't want to work
                    in this situation, do you?"

Let's look at questions 1, 2 and 3 first and discuss 4 later. Each
of the first three are questions you might find insulting and
deserving of a reactive answer. If you do reply in such a
manner, you will be doing exactly what your interviewer is
trying to make you do: get emotional, react negatively, and
disqualify yourself for the reason mentioned earlier. The
natural response is to defend your GPA, or your college, or your
choice of service. The way you must respond is not natural: you
must **agree** with the interviewer...up to a point. This is most
unexpected, and very much a sales technique on your part. The
first words out of your mouth must be positive:

A #1   "Yes, I know that 3.0 is not a perfect grade
                    point average. But I have to tell you, I got a lot
                    more out of college than just studying. I was
                    very involved in student government, as well
                    as other extra-curricular activities. As a
                    matter of fact, I worked 15 hours a week while
                    going through school and that took some time
                    away from my studies. But if I had to do it all
                    over again, I'd probably come out with just
                    about the same average, because I feel that
                    those other activities were very influential in
                    my personal development."

A #2   "Yes, I know may of the students at XYZ
                    College are there at their parents' expense,
                    and many of them might not take their studies
                    as seriously as they could. I do know that I was
                    not fully subsidized by my parents, and that I
                    paid for the majority of my college expenses by

working at the bookstore. At the same time, I feel that the college gave me a very good education and prepared me to get started in my military career, which was a different educational opportunity."

A #3   "You know, I think that the Air Force has some definite advantages over the Army. I've been in some of their officers' clubs and have seen how nice they are! I gave serious thought to going into the Air Force, and the Navy as well, for that matter, but they both wanted a much longer duty commitment from me, much longer than I wanted to lock myself into the service for, especially when the Army offered a shorter requirement. *For me,** the Army was the best situation."

These answers can easily be expanded to include other information of a positive nature. Do you see what you accomplish? You can really make points by being level-headed and considerate of the recruiter's viewpoint if you do not react to the questions in an argumentative manner. The ability to show that you can handle your emotions and act cooly in the face of these negative questions will also show the recruiter that you will not be shaken by them, and no more of that kind will be asked. Is this a game? Definitely!

Let's look at the fourth question, which is a negative one, but of a different type. The purpose of this question, as you might suspect, seeing it here in black and white, reading about it, is to shake you from your sense that this job may be the one you want. The natural way to answer this question is to agree with the recruiter who has, after all, painted a rather bleak picture of what you might expect in the organization. If you agree that you would not want to work under those conditions, you may just as well get up and leave the interview, because it is over as far as the interviewer is concerned. This type of approach is fairly common. Although the example is a fairly obvious demonstration, you can see how the question could be asked in a much more subtle way:

---

* There are those two little words again!

131

Q #5   "This sales job needs a person with some
              business experience. I don't think that you
              have enough to really be considered. Have you
              ever thought about a job in sales service?"

In either case, your answer must have the same quality of
thinking behind it:

A #4   "I don't expect the job I take to be easy. I have
              never seen a worthwhile goal that was
              attainable with just a little effort. I'm not
              afraid of hard work, if the effort will bring
              rewards. After all, I am in the position where
              I must prove myself and my abilities. Any new
              career position is not really different from any
              new military assignment in that respect. I do
              think that I am the best judge of my
              capabilities, and whether I want to perform
              the duties you have outlined. I would make
              that decision based on the offer your company
              makes to me, and where I feel the opportunity
              can lead me in terms of personal growth."

A #5   "I can understand that you might feel I have
              little experience in the business world. But we
              have been discussing many positive personal
              attributes I feel are strong indicators of my
              capability. I may be at a slight disadvantage
              initially, but I know that I learn quickly and I
              certainly have the desire to succeed. I believe
              that if I apply myself to any task at hand, I
              can make it happen. I am most interested in
              making that happen in a sales position."

If you are one of those people who has interviewed before
reading this Handbook, you may be saying to yourself, "Oh no,
that's the reason I blew that interview." Welcome to the club.

Negative interview questions can be a lot of fun if you know
what is going on and can respond to the situation with control
and maturity. They can be a disaster if you get upset and react
with a negative answer. You must answer positively and with
a subtle sense of humor. Above all, remember that the

interviewer has nothing against you personally. This is just another ploy in the bag of tricks to see what the real you is all about in a short, face-to-face interaction. It is a good trick to see how you may react in a stressful situation. Don't fall for it.

Pursuit of the employer should begin even before your initial interview takes place. You should not stop pursuing the employer until an offer or a notification of rejection is received, or another position is accepted (and there is a good case for continuing the pursuit even then). Pursuit of the employer is simply showing active interest in the recruiter, the organization, and the position to be filled.

This part of your interviewing will have a strong impact on the overall impression you are creating. As you read through these pointers and suggestions, it might occur to you that they would lose their effectiveness if everyone did them – if everyone took the advice that follows. The fact is, you'd be absolutely right. If everyone wrote a note to the recruiter to say thank you for the time spent and information received, the act would lose its meaning. It would become **expected** behavior. As it is, very few people will have the foresight to do such a thing, mostly because it has not been suggested to them, but also because they are lazy or think they have better things to do.

This is advice that will only serve those who are serious enough about their career search to put the suggestions in action, as we discussed in Section 9, "Attitude Check." My experience is that fewer than 25% of the people who know about these techniques actually follow through with them. So you see, there is little chance that these tips will become uniformly routine. Let's start off at the point where you have an interview scheduled.

The initial interview is where your foothold is established. It is here that you will be considered or not considered for further interviews. Once you are in the second interview, you will find the fatality rate much lower.

Let's say that you have an appointment scheduled with Mr. Jones from XYZ Inc. in about two weeks. Mr. Jones is coming all the way from San Francisco to interview you in New York at a military interviewing conference*. Once you have made the appointment, I'd strongly suggest you write the good Mr. Jones a short letter, including the fact that you are looking forward to the interview at such-and-such a time on such-and-such a date, and include a copy of your resume (with the proper objective on it, of course).

What this does is set you apart from the other 87 people he is going to interview in New York, and gives him some information he can digest and relate to your name. Little does he know that this is only the first in a chain of small but meaningful actions on your part. He doesn't have a chance.

The next real opportunity to show your interest is in the interview. Certainly, your enthusiasm and desire are going to be impressive, but there is one more step to go: **telling** Mr. Jones that you really like what you see and that you are interested. Now that doesn't sound all that unique, does it? I mean, if you are applying for a position, and it looks good to you, isn't it natural to tell the person you are talking with that you like the situation? Sure it is, but I have seen countless applicants interview for half an hour, finish the conversation and get up and leave the room without saying a word about being even interested in the opportunity. It might have been that they were so nervous and up-tight, and they were so happy that the interview was finally over, that they just wanted to get the hell out of there as fast as possible. That could have been a very normal reaction.

I'm saying that you should **never** do that. If you have gone to all the trouble and expense to interview with Mr. Jones (and he with you), you should **tell** him you are excited about the situation and thank him for his time as you leave. This gives him the information that if he decides to follow up on you, the odds are that you will have interest in his situation and he likes the assurance that he can count on you. Recruiters don't like to be rejected, either. Think about it. (There is one other thing you must do before you leave that room, and you will read about that in the section titled "What's the Next Step?")

---

* Unfortunately, it is not always possible to know the name of the person you will be talking with, particularly at a conference or job fair.

After you are finished with your interview, you should immediately write a note to Mr. Jones and tell him that you have interest in and are impressed with his company, and are looking forward to hearing from him again. It really doesn't matter what you write as long as it is short, sincere, pertinent, and conveys the idea that you are interested. I would recommend typing this note if you have the ability, but do not take the time to get to a typewriter if that is going to delay your letter getting off. The note should be short and to the point **unless** you feel that there are some important points needing clarification as a result of the interview.

A good example of this would be if you were asked to describe a leadership problem you had in the last five years. Perhaps you answered the question, but after the interview thought of a much better example. Well, maybe you have already blown it, but it is certainly worth the time and effort to express that new example on paper in the hopes it will repair any damage you may have done with the earlier (mediocre) example of leadership. Remember, you are going to want this note to be no more than a couple of paragraphs. Do not try to use it to correct **all** the errors you have made in your interview.

So here is Mr. Jones, returning to San Francisco after interviewing 88 people in New York. If he is typical of most recruiters, he goes to the office and one of the first things he does is open his mail. And there, in his in-basket, is a letter from you thanking him for his time and effort and expressing your interest in his program. If you don't think that is a welcome feeling, you have never traveled across the United States and, through the jetlag, interviewed 88 people. If you had a good interview, this note is a strong positive reinforcement of the recruiter's good judgement of you. On the other hand, it might do you absolutely no good. You might have been rejected on the spot in his mind. But if you were on the line, if you were in that gray zone, this little effort might just push you over the edge into the "consider" category.

It sure is worth the price of a stamp to send the letter, isn't it? There are some applicants I know who have so much faith in this little effort that they have their notes already written before the interview, stamped and addressed, ready to drop into the nearest mailbox as soon as they walk out of the interviewer's office.

You should continue these two processes (sending a note before, when possible, and after the meeting) in every step of the interviewing sequence. You can confirm the time and place of your second, third and fourth interviews, and send notes to each of the interviewers you encounter along the way. Show your enthusiasm, show your interest, show your organizational ability. It might pay big dividends.

There are other, more subtle, ways of pursuit. If you are expecting a telephone call and need to be somewhere else, make sure that whoever is going to be at home while you are out knows to expect the call. Example: Mr. Jones calls from San Francisco. If your spouse or parent answers the phone, Mr. Jones must explain who he is and why he is calling. No big problem, but if whoever answered the phone can look at a list by the telephone and say immediately, "Oh, Mr. Jones, John had to leave for a few hours. He'll be disappointed that he missed your call. He's been looking forward to talking to you at XYZ Inc. Can I take a message, and have him call you as soon as he returns?", don't you think that Mr. Jones is going to feel that you are interested, since even your family members know all about him and your interviewing?

Return calls immediately, even if you believe it is past the closing time of Mr. Jones' office. If he is typical of most recruiters, he is in his office well past five o'clock. Every hour you delay that return call takes something off the image you are trying to create. That recruiter is wondering what it is that you are doing that could possibly be as important as talking with him about further interviews with his company.

Where Mr. Jones, the initial recruiter, is responsible for coordinating the entire interviewing sequence, you should take the time to keep him informed, independent of the contact you have with other interviewers in the same organization. It might work like this: you interview with Mr. Jones. He refers you on to Mr. Smith, who refers you to Ms. Baker, who refers you on to Mr. Adams, who gives you an offer to go to work for the organization. If Mr. Jones is a recruiter, he may not hear much about you after he refers you to Mr. Smith. It is very positive reinforcement for Mr. Jones to receive a note or a phone call from you at each step, telling him what you have done. Tell him that the company opportunities are as exciting as he described (or even better), and that you are very grateful to him for giving you the opportunity to go through the interviews.

This is great strokes and job satisfaction for the recruiter, Mr. Jones, but what does this accomplish for you? It builds a relationship between you and Mr. Jones that might be needed to weather a storm. That is, if you are rejected for position A, he might be so impressed with you that he will refer you on to position B, which would mean an entirely new set of interviews with other members of the organization. Even if your interviewing is going well, it might make things go even better if Mr. Jones talks to Ms. Baker and asks how you are doing in your interviews with her, and remarks that you are certainly an interested applicant because of your contact with him. Use your judgement, based on Jones' personality and attitude toward you as a person.

In the section above, I refer constantly to a note or written communication. You should probably avoid use of the telephone with this technique. A telephone call can interrupt, cause inconvenience, and be a negative factor if Mr. Jones is in an interview or in a conference with his boss when you call. Try to use the note format; it can be read at the recruiter's convenience.

If your relationship develops to the extent that you feel you can call Jones after that next interview, do so, but only if you feel that your call will be catching Jones at a convenient time. If you have the kind of relationship that will enable you to make that call, you should be familiar enough with Jones' schedule through your exposure to him and his secretary - if he has one. Note that the Jones to Smith to Baker to Adams process might take three months or one day, depending on the situation, and you will adjust your contact with Jones accordingly.

If someone is dragging their heels in the process, you will need to resort to some attention-getters. Example: Mr. Smith says that Ms. Baker will call you in about ten days for another appointment. You send your note to Smith, thanking him for the interview and telling him that you are looking forward to talking to Ms. Baker in ten days. You then drop a note to Ms. Baker confirming the fact that you have talked to Smith and that you are expecting her call. She doesn't call. On the eighth or ninth day, call her office and talk with her, if possible, explaining that you might have missed her call because you were leading the local Scout troop through Brenner Pass (or whatever), and that you are still very interested in talking with her. If that doesn't get results in a reasonable period of time, you might try calling Smith or recruiter Jones and telling him that you have come to a dead end, and you really want to continue interviewing with his company.

Be nice. Don't assume that the mistake is on the part of the organization. You are treading on thin ice and have to assume that since they have not followed through immediately, they may be less than 100% convinced that they want you to join their ranks. Don't jump onto that thin ice with a sledge hammer, swinging it about and making accusations that someone said they would call and did not. Remember what we said about recruiters and interviewers having some other responsibilities that might take priority. If you are not getting through to the people you need to talk to, stop calling and send a Western Union Mailgram. Just as their advertisements say, it is hard for anyone to ignore a special communication of that kind. They get read first.

If you are given an application form to fill out and send to the recruiter, follow the steps outlined in the section "The Application Form," and then show your concern by sending it to the recruiter by Certified or Registered Mail. You will have a record of sending it and you know it has a better chance of getting to its destination. That is professional, that is efficient, and that is very businesslike. (That is also very expensive, but you will not have many to mail to the recruiters you talk to.) Any important documents or correspondence (your transcript of college courses, your letter of acceptance, your aptitude tests) should be sent in the same manner.

These are all small points relative to the importance of your interview content, and the effort you have or will put into self-analysis and question or answer preparation. Do not be misled into thinking that these tips can take the place of good, solid interviewing techniques. They **might** help you out of a tight spot and give you that extra edge when you need it. Be different and try them.

Very few offers are given to applicants in their first interview. Indeed, you may not want an offer to be made at that point because you would not normally have enough information in one short interview to fully evaluate the opportunity. The same goes for the recruiter - generally, more opinions from other managers are necessary to make an evaluation that would result in an offer.

What usually occurs is a series of interviews, the number of which is different from organization to organization. If your first interview was with a **recruiter**, you would be referred on to a manager for your next interview, with the recruiter's stamp of approval. For the purposes of this section, we will use the example of our recruiter, Mr. Jones, who is referring you on to a field sales manager, Mr. Smith. The names have been changed to protect the innocent.

You have had a good interview with the recruiter Jones and have an appointment to meet with Smith, the regional sales manager, for a second interview later on in the week. With that information duly noted on your calendar, the first thing you should do is try and find out as much as you can about Smith: where he lives, what his area of responsibility is, whether he has the authority to hire you, and any other information you can get. This information could come from the recruiter **if you ask correctly** or from whomever has set up the appointment, such as the employment agency or placement office.

Why the need for all this information? Because your interview with Smith is going to be different in many respects from your initial interview with recruiter Jones. The sales manager will be looking for different things, and taking certain other things for granted. Remember the interview with Jones? You stressed communication ability, direction, ambition, management potential, past performance in leadership positions, educational qualifications. The recruiter was looking for all those things, all projected against an image of what the managers are looking for in a new sales employee. In this respect, the recruiter Jones is much the same as an agent in a good employment agency: now that he has found a good person with the basic qualifications, he must match the applicant with the needs in the organization. He will take into consideration all he knows about the personality and likes/dislikes of Smith. It is Smith, however, who will be looking at you in the role of a sales representative in his region, contrasting you to what he knows are the qualities needed to be a successful sales representative in the field. There are almost always some differences, some minor, some not, between what the recruiters see as essential and what the managers see as essential. You must look at this second interview with the same degree of empathy for the employer that you used in the initial interview with the recruiter. Above all, you should realize that the sales manager will look at you in a different light than the recruiter did. Why? Because you are going to be **working for the sales manager**, not for the recruiter.

A good example of the different perspectives you might expect in the two interviews would be those in a discussion of your management potential. In the first interview, since you knew that you were interviewing for a position that would lead you to management through the sales path, you stressed your management capabilities to as great an extent as possible. Your discussion of your sales **ability** was not as detailed. You talked about your **potential** sales capability. That is what the recruiter was interested in: a developmental sales type.

Now you are talking to Jim Smith, the sales manager. Does he have any interest in your desire to be a manager? Sure he does, but to a much lesser degree than he has interest in your desire to be a good sales representative, one who is going to go out there and work as if there was no tomorrow, to make good sales figures in his region and make him look like a hero so he can get promoted. He is just not going to be turned on by you talking at great length about how you want to be a sales manager. His attitude, and rightly so, is going to be "First you show me that you can be an outstanding sales rep, and then we'll talk about you being a sales manager."

To draw a parallel: if you were going out on a first date with someone you liked very much, you would not talk about what kind of silverware pattern you wanted to have and what style of wedding you were planning. That would be jumping too many steps, and the reaction would be: Whoa! Wait a minute. Your friend would have many questions to ask and things to see before making those kinds of decisions. Talking about strong management goals to a sales manager can be negative in the same manner. You are expecting too much too soon, and many sales reps leave the organization because they cannot realize their management goals within a pre-determined period of time. Looking at the interview from the manager's point of view can be very important. Your answers should differ slightly from the answers you gave the original recruiter. If not different in content, then in emphasis.

Another fact to take into consideration in this second interview: Mr. Smith is not a recruiter, he is a sales manager. He has probably not developed the interviewing techniques a recruiter has through daily exposure to the interviewing process or through specialized training. The manager's questions will be more applicable to the specific job at hand, and he will **react** to your answers differently. His questions might be more concerned about facts than theories: what you have sold, what are your demonstrated abilities in leadership, what other sales jobs you are considering, who do you know in sales, how you will fit into his group. Understand that this does not necessarily make the interview harder. In fact, since the recruiter has screened you, the manager might take for granted that you are an acceptable applicant and have a much easier attitude about interviewing you.

In our example here, where you have been scheduled to meet with a sales manager, there is one other major difference between the two interviews: Mr. Jones, the recruiter, will make a decision in a very **objective** manner. He will normally not even see you again after you start working for the organization. Certainly the odds of your ever working for him in that company are long odds at best. His long-term interest in you is minor. On the other hand, Mr. Smith, the sales manager, is going to be in direct contact with you for as long as he is in the position of being your manager. He has a much more subjective, emotional decision to make about you. He is visualizing you in the role of working for and with him, joining the team. Use his **subjectivity** to your advantage. Use some of the information you have collected about his hobbies, family and other interests. Just about everyone's office has some indicator of hobbies, family or other interests you could use to develop rapport. Work with the manager's personality. Don't make the interview a pure business situation, but enjoy yourself and let your own personality come across as well.

In the case of a sales career, the personality of the sales representative is extremely important in conducting daily business. Don't hide yours in a box, let it flow naturally. Some civilians hold a stereotype of military people. If you come off in an interview as if you are reporting to a new duty station, all stiff and structured, you confirm this suspicion.

The third interview in our example might be a field trip. You will be scheduled to spend a day in the field with a sales representative or a sales manager (in this case, but it could be at a manufacturing plant or a corporate headquarters if you were interviewing for other kinds of positions), doing the same job you will be doing if you are finally hired by the company. In this interview, you are going to find out many of the things that will influence your decision to either take the job or not, if it is offered. In fact, I don't see how any person can accept a career position without this valuable input. It is here that you are going to find out what the job is really like. Up until now, it has been all talk.

As in every case, you must find out all you can about the person with whom you will be interviewing. Yes, interviewing! This is an interview and a very real one, as is any contact you have with the employing organization. The temptation is certainly there to let your guard down. A sharp company is going to set up a day to be spent in the field accompanied by someone you can relate to. Perhaps you will have graduated from the same college, be about the same age, or maybe served some time in the same branch of the military. This has not happened by accident. It is meant to put you at ease and give you something to talk about as you spend the day calling on customers, getting in and out of the car, eating lunch and driving from account to account. Don't be dumb and start to think that this person is your friend. Your "friend's" interests are in getting promoted,

146

and there is motivation to tell the manager about any unsavory details of your life discovered during your time together - details which could disqualify you from consideration. Look at your field trip as a day-long interview. That's a bit more scary, isn't it? That should keep you alert.

Here is your opportunity to ask lots of questions at the operational level. Do it. Show your interest and your concern by finding out exactly how the organization works. The detail will help you make up your mind if you are given an offer and will have the additional benefit of orienting you when and if you do take the job. This field day is such an important step in forming your opinion of the opportunity that I am amazed that companies do not put more emphasis on the selection of the representatives who are to conduct the "interview." Sometimes it is a random selection, more determined by who is available than by who will give the best impression and do the best job of interviewing. Of course, the employer does not want to give a false impression, but there is no excuse for sending a prospective employee out in the field with an incompetent. It happens, and more often than the company president knows about, that's for sure. For that reason, you should not base your entire decision on this one day or on this one person, but rather on the picture you see after talking with the whole spectrum of individuals you meet.

WARNING

There is a great temptation to assume that the first person you talked to has given interview notes to the second, and that person to the third. You might assume they are discussing your interview in detail and are making a nice file filled with notes on your background. Let me assure you, that is most likely not the case. As far as the second person in the sequence knows, you are a candidate the first interviewer feels is right for the open position, and that is it. Occasionally, some evaluation might pass between interviewers but, as we will see, it is best to assume that none has, even though the temptation is to refer back to the previous interview:

Q: "Why did you decide to go into the service after college rather than start a career in business?"

A: "As I told Mr. Jones, I decided that I wanted to go into Officer's Training School."

Hey, Smith doesn't care what you told Jones! He wants to hear it for himself. More importantly, you are falling into a common trap, and there are lots of bones in the bottom of this pit: as you go up the ladder of interviews, you are asked the **same questions** over and over and over again. There are just so many questions that can be asked of a person, and you will hear several more than once. AND YOU GET LAZY. You assume that the first person has given a tape recording of your interview to the second and the third and fourth person you will be talking with and your answers get more and more shallow as you go higher and higher on the ladder. An answer which has good depth in the initial interview, covers all the bases and takes one and a half minutes to present becomes, in the fourth interview, a virtual yes/no answer, incomplete and shallow. After all, it gets boring to say all those things over and over again, right? WRONG! The fourth person is probably the most important, is ready to make an offer to you based on the glowing comments received from your previous interviewers, and you come across like a wet dishrag. Forget it! You have lost the whole game because you have not stayed alert, treating each interview as if it was the first and most important. Leave nothing out of your answers from interview to interview. If anything, you should be adding to your answers, making them more complete and in-depth. Do yourself a favor and don't allow this "I've got it made" or "I've said this before" attitude to affect your answers.

I cannot emphasize enough how important it is to keep your guard up, and keep your answers complete, no matter how casual the situation seems. True story: there is a well-known company, which will remain nameless for obvious reasons, which brings applicants into its headquarters from all over the U.S. for a series of interviews. The following occurs: a meeting at the airport, transport to the hotel, settle in, taken out to a nice dinner, perhaps to a ballgame or concert, and back to the hotel. Red carpet treatment.

The applicant goes to bed, big day ahead, and the employee goes home to write a mandatory report discussing the points evaluated during the "interview" with the unsuspecting

148

applicant and has the report on the recruiter's desk before the applicant reports for the first interview the next morning. Any areas considered to be shortcomings are thoroughly investigated and probed during the day's conversation. Pretty slick, isn't it? And if you were not aware of what was happening in the situation, you could be sunk before the boat race even started. You thought that this company representative was a "friend."

After two or three interviews, you should be asking the interviewer about an offer. You have seen the job firsthand, done your research and asked all the questions you are going to ask. So that you can easily refer back to the offer-making process in this Handbook, I'm going to include it in a separate section later on. It is the most interesting part of the whole game. But before we go on, let's review the very important information covered in this follow-up interviewing section.

There is a big difference in interviewing with a recruiter and interviewing with a manager interviewer. They look for different things and from different viewpoints. You should never be lulled into assuming that the first interviewer talks to subsequent interviewers and, most importantly, you should keep a high level of information and depth in your answers as you go up the interviewing ladder. Resist the temptation to let up because things are going well. Finally, you should keep your guard up against seemingly casual situations every time you have contact with **any** representative of the potential employer.

If you are just returning from an overseas tour, here is some valuable advice which might save you lots of trouble and embarrassment.

Most organizations which request that you appear for a second interview will pay your expenses from point A to point B, including meals, parking, hotel rooms, airfare, rental car or mileage. However, not all organizations are prepared to pay for those expenses in advance. This is not because they distrust you, but rather their accounting mechanisms function poorly on a pre-paid basis. Most can provide airplane tickets (which could be waiting for you at the airport or provided by travel agencies), but pre-paying a room or a rental car is a lot more work. Here is what you need to do:

When asked to go to Remote Location, Louisiana, (population 3,100) to interview at the Toothpick Factory, express your joy at being offered such a unique opportunity and keep your mouth shut. Whoever is coordinating the trip, often the recruiter, will go on to explain the arrangements. In doing so, you will be told, usually, that the company will pay for your expenses. Fine, you don't need to ask. If it isn't clear, simply ask: "Should I go ahead and buy the tickets and save my receipts for reimbursement?" If the answer is no, and you are on your own expenses (a highly unlikely event with major companies), you might need to reconsider how badly you want that particular offer before investing your own money.

Most of the time, the company will pay for all your expenses, but be prepared to front some major dollars before you are reimbursed. If you do not already have a major credit card or two, you'd be very wise to look into the process for getting one - very handy for this kind of travel.

If you find yourself in the situation of needing to get to a location for a follow-up interview and have no credit means to do so (flights can max out anyone's plastic), I'd suggest calling the **recruiter** to discuss the problem. Recruiters understand that you are not American Express. They deal with first-time candidates frequently. But the national sales manager who asked you to make the trip that you can't afford is accustomed to dealing with business accounts covered by other than individuals' finances.

Keep careful records and **all** receipts. You don't want to make any money on your expense account, but there is no need to lose any, either. Be clear about where you should send these reports. Expect a two- to four-week delay before getting reimbursed, and keep a photocopy of your expenses in case the originals get lost in the accounting department.

For this section, I am going to assume that you are applying for a position with a major national or regional organization. While the following information is also applicable to smaller companies, it is particularly sensitive when you are dealing with larger companies.

We are going to cover three areas that are responsible for the rejection of otherwise qualified applicants time after time. You will learn to avoid giving finite, or direct, answers to these three classifications of questions. I've called them "the three finites." Avoid them at all costs, but if you are confronted with one of them directly and an answer is demanded, try to generalize the answer to minimize the harm that you might do. If you are to receive the maximum number of **offers** in your interviewing, you must pay attention to these areas.

THE FIRST: the geographical location where you want to work. We have touched on this subject elsewhere, but it was not presented as strongly as in this section. If you are interviewing with an organization which has operations in each of the eleven western states and are asked where you want to work, your answer **must be** that it is relatively unimportant to you where you are located. There are two different points of view from which to look at this: the recruiter's and the applicant's.

Let's look at the recruiter's thoughts on the subject. If a recruiter has interest in a particular applicant and asks the question, "Where do you want to work?" this is really asking a misleading question. What is being asked translates to: "Does it matter to you where you work?" or "Where will you **not** work?" And, if you answer this question: "I want to work in San Francisco," this answer is interpreted as meaning a) that you will go to work elsewhere only if you don't have an offer for San Francisco, b) if you want to work only in San Francisco, you might have to take a job elsewhere but you are not going to be happy and will shop around for a S.F. job as soon as you can, c) if you start with this company in San Francisco (there might be a job there), at some point you are going to have to be relocated and the odds are very good that you will quit your job rather than move out, and d) if this company's headquarters are in New York, the possibility of you moving there during your career is very great. If you won't move, why hire you in the first place?

Now you might say that is a whole lot of overreaction to the simple answer "I want to work in San Francisco." Perhaps it is, but remember the recruiter is responsible for recruiting good people who will stay with the organization for careers and searches for reasons why you should **not** be hired rather than reasons why you should be.

Let's look at the situation from the applicant's point of view. First, let's be honest. What is more important to you right now? The **location** in which you are working, or the **opportunity** in which you are to work. If location is extremely important to you, I recommend that you seek employment with an organization that will not take you from that location. Don't apply to national organizations. You are wasting everyone's time. You really need to be honest with yourself on this subject at this point in your life. If you make the mistake of giving priority to the wrong factor, you are going to be unhappy in the future. You have no business looking at a national organization if you want only to work in one city. There are few career positions which will allow you to stay there for the rest of your career. Why go to work in that sort of situation if you are unwilling to relocate! If you do a good job and are promoted (read "relocated") and quit rather than move, you are back to square one having gained little more than a couple of years worth of paychecks. You could have probably made more money as a bartender. Ideally, the most important thing to you right now should be getting involved in a worthwhile and challenging situation regardless of where that might be. I mean, that is the way it **should be**!

As usual, the realistic situation is probably somewhere in between the two extremes. You are open to many locations depending upon the opportunity, but you have some preferences. Doesn't that seem reasonable? "I will work anywhere, but my preference is for San Francisco." That might seem reasonable to you and your family, but you are still telling the recruiter that you really have a desire to work in S.F. All thoughts a,b,c, and d above are probably equally applicable to the applicant who has a "preference" as to the applicant who says working in that location is the only choice. The solution to the situation is to give an open answer, omitting your preferences until asked for them.

Q: "Where do you want to work?"

A: "It really matters little to me where I start to work. I know that I will probably be relocated from there in a couple of years anyway. What is important to me is the opportunity to learn and grow in the position."

Q: "Do you have a preference as to location?"

A: "Yes, I do (every intelligent person is going to have at least a preference, and you have been put on the spot to name it). I have lived all over the world in the military, and was once stationed in San Francisco. If all other factors were equal, I'd choose that place to start in."

Perhaps the total truth of the matter might be that your personal priority is working and living in San Francisco. Why are you going to all the trouble to tell people that you really do not care where you work? The answer is simple: you want to collect offers. You want to avoid going into interviews and closing the door on every opportunity that is not in San Francisco. If you effectively turn down a potential offer for Los Angeles by simply coming on too strongly for another location, you might find that Los Angeles starts looking very good six weeks down the road when your interviews have produced zero offers for San Francisco. There are only a very few situations where you could reopen that closed door.

As a military leader, something that makes you attractive to major employers is the fact that you have traveled and relocated during several moves in your career. You are familiar with the problems of relocation, and your experience in handling those problems (possibly with your family) are valuable experiences that your fresh college graduate competition has probably yet to learn. Recruiters need to know that. Use your asset.

The first finite, then, is the finite of location. It is super-sensitive. You generally want to avoid giving an answer which would indicate an absolute preference for a specific location, even if you knew that a recruiter had an opening there. A good recruiter for an organization which covers a large geographical area is looking for individuals who will be willing and able to move freely within the boundaries, whenever and wherever that person is needed to do a job.

THE SECOND: money - money you want as a starting salary, money you hope to make in the future. The subject is a very emotional one. You want to make as much money as you can, right? As with the thoughts on relocation, I am going to tell you some things you might not have thought of up to now, so keep your mind open.

I would say that the vast majority of all positions you will interview for, whether those positions are with major companies or with smaller organizations, have a set salary range allowable for entry-level jobs. Typical for a larger organization is a scale on which each and every job performed in the company is given a dollar bracket in accordance with the responsibilities and qualifications set out in the job description.

Under this system, a janitor might be a range level 1 and the president might be a 20. You can assume that the range for your job is going to fall somewhere in between the two, let's say a seven. Within that pay range of a seven, there is generally some flexibility. But unless you have some directly applicable experience, you will start toward the bottom of the range. It is not very often that a recruiter or interviewer is given the authority to "pay whatever is necessary to get him," and that would certainly be a rare occurrence at the entry level. My point is that unless you had some kind of civilian work experience before the military or have done directly related work while in the military, you are not in a position to do a whole lot of negotiating when you apply to corporations. Let's examine a sample set of questions about money.

Q:	"How much money do you want to make in this position, Jack?"
A:	"I want to make $24,000 per year."
Q:	"That is unfortunate because all I am allowed to offer at this level is $21,000."
A:	"Oh, that's all right. I guess that money isn't all that important. I'd work for $21,000."

What is gained by stating a finite sum of $24,000? Absolutely nothing! As a matter of fact, you could say some harm was done. That exchange does not exactly present a clear picture of decisive thought, does it? Additionally, the whole interview could have very easily been blown. Most recruiters would not offer the fact that they could only give a person of those qualifications $21,000. Most would likely only file the fact and then bounce the applicant out of consideration when the next person came along with similar qualifications, but who answered the question like this:

A:	"It is really a consideration of secondary priority to me right now. I assure you I want to make a good salary, but right now I am much more concerned with the factors of job responsibility and of 'chemistry' with your company. Without those, even a large salary would not attract me to the position."

Again, let's look at the **ideal** attitude. If you are starting out in your career, money should not be an overwhelming consideration. You need the opportunity to grow, to learn some skills that will bring you to a higher level of responsibility, to prove your abilities so you will be worth more in the marketplace. That is the open attitude the ideal applicant demonstrates by putting money in a proper (lower) perspective.

You have gained some good experience in the military. You came in at a low level and progressed through the ranks. If you were a Lieutenant, you gained experience and by the time you made Captain you knew enough about what was going on to be worth more money to the military. That is why you are paid

157

more at higher ranks. The same is true of civilian employment. You are starting out near the bottom and progress through the ranks of experience and are paid for it. There is one point that is different. You have worked and been paid for your experience in a field (the military) that is **not** related to the civilian employment scene. You are applying for an entry-level position and making more than other entry-level applicants. It is going to be even more difficult for you to accept the "ideal" attitude described previously. Work at it. Do not assume that you will be paid more because of your years of military experience.

That "ideal" attitude (money doesn't matter) is perhaps a bit hard for you to adopt. You are looking for a good developmental opportunity that pays you as much as possible. The truth of the matter is that an entry-level applicant can probably live on any salary that a respectable organization is prepared to offer, unless you have some very expensive vices. Your organization will have an interest in you and does not want to see you going around with holes in your shoes. There are exceptions, but you will eventually eliminate those from your consideration. Again, you must fall back on the philosophy of interviewing for offers rather than for a specific ideal. Theoretically, all organizations you interview with will be in competition with each other for your talents. Company A will not offer you $25,000 while Company B offers you only $15,000 - not if they want to compete for the same caliber of applicant.

Do **not** ask for a salary which is 30% higher than you hope for and expect the company to offer you less than your stated amount. That is playing salary roulette and you aren't able to afford the odds. A recruiter would much rather reject you as having "money goals which cannot be realized in our company" than give you an offer you will turn down because the money he is allowed to offer is not enough.

Even on your application, the blank that asks for "salary expected" should be completed with the word **"open."** You might be pleasantly surprised and find that the salaries you have to decide upon will be higher than you expected. Then, taking all the facts into consideration, you can make your career choice as a business decision.

There are situations where you will be put on the spot. How do you know what to say? It is equally harmful to guess low as it is to guess high.

Q: "How much money do you want to make in this position?"

A: "It is really a consideration of secondary priority with me. I assure you . . . etc."

Q: "That's fine. I am really glad to hear that you are open on salary. That makes my job easier. But I really want to know if we are in the same ballpark here. There is no reason to continue our interviews and fly you to New York if you want more money than we can offer."

While it might seem that you are in a corner and must give a finite answer of X dollars, you do have one last card to play - the "range" card. If you are working with someone who might have salary information, you should have picked up that information before you entered the interview anyway. That person might have been at your employment agency (where, I assure you, the starting salary range is well known), or at your placement office, or perhaps even the uncle who set up the interview. All you need is a range. If your information says that the salary is around $26,000, give a range of $25 - 28,000 in answer to the question, but repeat that there are other important considerations in making your decision. If you do not have a range, you are up the well-known creek and need to make a guess. Take a wider range than you would if you had salary information, say $24 - 29,000, and put more emphasis on the fact that it is of secondary importance. Keep your fingers crossed. Remember that few recruiters are involved in the actual salary negotiations. That is usually between you and the hiring manager. The recruiter is there to screen you out if your sights are set unrealistically high.

Finally, you should realize that there is a big difference between starting salary and first year income. When you take into consideration that there will probably be raises, fringe benefits, possibly promotions, bonuses, and maybe even an expense account and company-paid automobile involved in your income, the base salary could really fade in importance. All that comes out in the offer.

Now, having read the sections above about money and location, if you were asked the question:

Q: "If you were offered a position in Chicago at $24,000 and a position in New Orleans at a salary of $29,000, which would you take?"

How should you answer? If I have done my job well, your answer should go something like:

A: "I cannot tell you. There are so many other factors to be considered that I'd need more information before I could say."

That is the attitude of the person who is seeking a **career** rather than just a **job**. That is the person who is going to be taken seriously by the interviewer. It demonstrates common sense and smarts not to make a decision of this importance based on just salary and location even though they are two very important and emotional variables.

The second of the three finites is the money factor. You now know about it.

THE THIRD: Future position - where you are going to be in that distant or not-so-distant future in the organization.

The position you are interviewing for at the moment should be firmly set in your mind. But what about where you hope to go once you are in the race? This subject will come up most often in the form of a misleading question:

Q: "What do you see yourself doing in ten years?"

As in the case of the other two finites we discussed, the smart thing to do is to answer in an indefinite manner, giving enough substance to show direction but not finite enough to show unrealistic goals or goals based on ignorance, which will eliminate you from consideration. There is a different way to ask this same misleading question about the future:

160

Q: "What kind of money do you see yourself
            making in ten years?"

Look, there is no way you are able to answer either of those
questions in finite terms. You have little or no exposure to the
natural progression of even the **average** career growth of a
person in that organization. And God knows you, after your
outstanding military career, with all the leadership
experience that has brought, can hardly think of yourself as
average, right? Let's look at an honest and realistic answer to
the questions above:

A: "At this point I couldn't give you the exact
            position I want to reach in ten years or even
            what salary I would hope to be making. (To
            take it another step and show some direction:)
            I would hope to progress commensurate with
            my abilities. I would say that in ten years I
            would have plenty of time to show that I am
            an exceptional employee, and that I would
            have been given the opportunity to manage a
            certain portion of the company's business and
            perhaps even some other employees. But
            exactly where that level would be is a function
            of the company's attitude toward my successes,
            the   availability   of   openings   in   the
            management structure, and the overall success
            of the organization."

You might reread the section in this Handbook on "Misleading
Questions" if you cannot see why the question is a misleading
one.

It has been suggested that you might answer the question in
this manner:

A: "In ten years I'd like to be in mid-
            management."

To the uninformed, that might appear to be an acceptable
answer. But you know what's coming, don't you?

Q: "Define mid-management. How many employees, how large a budget, what are you responsible for as a mid-manager?"

Unless you are intimately involved with that organization, you will not be able to answer that question accurately. You are talking from a position of ignorance in spite of the fact that "mid-management" sounds nice as a ten-year goal. What this boils down to is that you should not throw terms around for which there are no concrete definitions applicable in all situations. You tell me: how large is a territory? How long does it take to become a plant manager? How many employees does an office manager supervise? Does a regional manager outrank a division manager? What is a top-level manager? The answer to all those questions is: it depends. It depends upon the particular organization and how it is structured. In Company A it might be a very realistic goal to become a plant manager in five years and be making $60,000. In Company B, that position is reserved for people with much more experience and will command a salary of $150,000 per year. There isn't a common denominator for the term "plant manager" or any of the classifications above in terms of salary, experience or scope. You cannot answer the question unless you know how the term applies to the organization you are talking with.

As a military person, you are particularly susceptible to this ploy. You are familiar with talking about commonality in job descriptions. A platoon leader in the Army Infantry is a Second or First Lieutenant, who has responsibility for (ideally) 40 soldiers including one senior non-commissioned officer and a number of junior non-coms. If you walk up to any Infantry officer, regardless of where he is working in the Army, he can tell you that information. But if you walk up to a production manager, he will be at a loss to define "production manager" in the same manner. The job is different from industry to industry, company to company, and even from plant to plant within a given organization.

Don't try to project where you will be financially or in career position within the organization in definite terms. Avoid using terms that sound good, but at second glance are shown to be meaningless unless you are in the position to speak with authority on how they relate to the job and organization at hand.

There you have it. The three finites. Just knowing what to look out for in those three areas will enable you to answer questions successfully, which would otherwise eliminate you or contribute to the negative side of the balance in an interview. Your answers will be "I can't tell you because . . .," with good solid thinking and demonstration of direction, but without commitment to an arbitrary, instantaneously contrived goal of location, money, or future position.

That is the question. The answer is much less clear.

I am treading on shaky ground here, mostly because the variables that seem to determine the use of the recruiter's first name are personal. There are times when the use of the recruiter's first name could be out of place, and times when the use of the title Mr., Mrs., or Ms. would be equally so. To make that determination, it is necessary to take several facts into consideration.

First, how much older than you is this person? I think it is really an artificial situation when someone my own age, or in the same range at least, calls me "Mr." If you are not sure if you should use the first name, use the title.

Second, what is this person calling you? If the recruiter starts off by calling you by your first name, you should follow suit. The trend has been toward the informal in the interviewing room – not to be confused with casual. The intimidating interviewer is out of touch with the marketplace and the current state of the interviewing art. Intimidation places people in a reactive state and tends to make them protect themselves and their pride rather than open up and talk about themselves. As a result, the interviewer who threatens candidates ends up knowing lots less about them than the one who is relaxed and inviting. There are still some power-hungry, immature recruiters out there, but they are less successful than the others.

Third, what is this person's personality and attitude? If it seems stiff and aloof, you might use the title until the atmosphere loosens up (if it loosens up). Use the title if you are unsure.

What are we trying to accomplish here? It is important to give the feeling that you are sure of yourself and your capabilities. In short, that you are the recruiter's equal . . . potentially, anyway. You will not accomplish this if every other word out of your mouth is "Sir" or "Mr." or "Ms." You are subordinating yourself. And if that interviewer is not much older than you, it seems even more preposterous.

Military people are familiar and comfortable with this subordination, and it is expected in the military, but not in employment interviewing situations. Go back to the objections managers sometimes raise when confronted with a military person for employment - "too structured, too stiff." Sir-ing everyone and everything in the room does little to change this image. There are people who are just not going to be comfortable calling their interviewer anything but "Sir" or by their title. If it makes that much difference and you **know** it will affect your interviewing to refrain from doing so, you should use the title, but infrequently.

Let's take a different tack. Unless you have been totally isolated from society for the past ten years, you realize that the title of this chapter could as easily have been "To Call her Jill or Ms. Jones." Some male military leaders will have a real problem being interviewed by a woman, particularly a younger woman, to say nothing of the issue of being on a first-name basis with her. The military is overwhelmingly a male-dominated, albeit artificial, society where opportunities for professional interaction between men and women are rare. Many recruiters are female – be ready for it, and be ready to be informal.

If you are a person who can use your interviewer's first name, do so. And do so frequently. A favorite question at the end of an interview where the applicant has not used my name, is to ask what my name is. This generally brings about a discussion of nervousness and uncertainty of whether the first name should be used. The use of my first name has never turned me off. It has always shown me that those applicants do have confidence in themselves. But I'm a first name kind of guy.

The woman who interviews for positions of a career nature could be frustrated in her efforts unless she opens her eyes to some facts of discrimination. As you read this section, you may say to yourself, "It's not fair." You may be justified in this feeling, but as in many other life experiences, fairness is not always determined by the referee, but by the players of the game. Before we discuss the unhappy side of the situation, let's look on the positive side.

The simple fact that I have made an effort to "neutralize" this Handbook by dropping most all references to recruiters, applicants, etc. as masculine is a reflection of the change in attitude towards women in the job marketplace. Just look around you in the military. Women **are** competing with men for responsible positions in areas previously considered for men only. Likewise, any employer who passes up the opportunity to interview a qualified woman to fill a position is passing up an increasingly significant portion of the work force in almost every profession

Indeed, you might find that your gender is an advantage as you start your interviewing. Some organizations are trying to make up for lost time, and are perhaps giving preference to the female applicant in some situations. (Note the wording of that last sentence. No lawsuits for me, thank you.) Others are openly encouraging the application of women under the pressure of EEO regulations and Affirmative Action programs (which many companies have willingly made a part of their culture).

Fine. These are possibly positive situations for you. But the fact remains that some employment agencies and employers will still ask how fast you can type while handing you the application form. I am not knocking the secretaries of the world. In fact, they could probably run the majority of the country's businesses as well as (and in some cases I've seen, better than) their collective bosses. But generally, a secretarial position does not have the necessary upward mobility to achieve your long term managerial goals.

If you are looking for a career in marketing, and are offered a job as a secretary, you have a right to protest, even in the face of the argument that a job is a job. A parallel would be an accountant looking for a finance career but being offered a job as a bank teller. A teller has a responsible job, but the tellers who get to the decision-making level of their banks are very few and very far between. So it is with the secretarial ranks - essential to the business world, but on a different track than the management-oriented career applicant. You should never confuse the two. Even though you might have an excellent assortment of secretarial skills, you have set your goal as a managerial applicant and to accept a position as a secretary demonstrates non-managerial direction (even though a good secretary might make more money than an entry-level college graduate in a management training program).

It may not be fair for a recruiter to brand you as a non-managerial applicant in the future just because you have gone in a non-managerial direction, but it is a fact of life. Women and men both have extreme difficulty convincing recruiters that they want to / can be in a management position if they have detoured off that path and worked as a non-managerial employee. As you read in the section, "Direction," establishing yourself as a managerial applicant can be a most important factor in your career search.

Let's look at some other snags. A woman is confronted with a corporation that is considering her for a position. Her marital status (an illegal topic of discussion) is sometimes a determining factor in the consideration for hiring. Why? Basic economics. If the organization hires her, she will be in competition for promotions. Let's say she does produce results and is offered a promotion in a location where there is an opening. If she is unwilling to relocate because her husband is employed in a career position at their home location, the organization must take the loss. Sometimes it is in the form of losing the employee, if not in the physical sense, perhaps in the motivational sense.

We said early on that in most large organizations, "management" equals "relocation," and if the employee is unwilling to relocate, the promotion goes to someone who is willing to make that move. Fair? Perhaps not. But it is only in rare cases that husband and wife are willing to live apart for career reasons, and historically the husband's position takes priority (perhaps because it tends to be the higher paid). I'm not going to get into a discussion of the sociological injustices. I'm just going to ask you to look at the married women you know and see if that statement applies to them. We discussed all of this in the Married vs. Single Status section.

The basic problem may never be resolved. If a woman is married and works in a managerial position for a major organization, a conflict eventually comes into focus: what has the greater priority, the marriage or the promotion?

I am well aware that many women who read this Handbook will not be married. The trend toward later marriages, the desire to become established in a business or career, and other factors that increase the numbers of single women on the job market, are not lost on recruiters. The single or unattached woman has great opportunity in business and other fields which demand relocation and full-time commitment. But the problem doesn't go away. Eventually, odds are you will probably want to marry. If your goal is to become a top manager of a major national organization, your path will be much easier if you stay single or marry an extremely mature, understanding man with a mobile career of his own.

If you face the decision head-on and decide the career you should seek is one not involving relocation, you should look into significant careers that do not require it. As we stated in the section on Large and Small Organizations, regional companies have plenty to offer. There are employment situations, such as in stock brokerages, which, although large in scope, do not **want** you to relocate once you have established yourself. Look at them and other related businesses: insurance companies, real estate brokerages, large regional retail organizations, anywhere you are expected to build a large personal client base.

If you live in an area where there are many corporate headquarters, careers that tend to be located in those buildings are worthy of investigation if you are qualified. Marketing management, human resources management, corporate level financial positions and other, generally, staff management careers, have headquarters locations. Moving from corporation to corporation, your career could develop in a different way than upward in one corporation. Expect a lot of competition for these jobs.

I have not attempted to paint a rosy picture in this section, nor have I colored it dark*. Employment opportunities for women have made great gains. The future holds even more promise as employers rid themselves of prejudices and are witness to the fact that women can do the job as well as men. Eventually, dual careers will have to be accommodated by corporations. Walk into the employment situation with your eyes open to your assets and limitations, sociological or personal, and compete.

---

* Over the last few years, I have heard from several women who have told me that they thought it took a lot of, well, nerve, to write this section. In almost every case they went on to thank me for the straightforward approach. The information comes from my personal observation of what is going on around me, not from newspapers and magazine articles. It's not a popular subject because there is a lot of unfairness in it.

There are books upon books filled with excellent stories of people who have successfully gone into business for themselves. Indeed, self-employment is a critical part of our system of free enterprise. All it takes is an idea, perseverance, some capital and (this is the part that impacts the interview) knowledge of how a business really works.

I feel it is short-sighted to go into an interview with any organization and tell the recruiter that you want to operate your own business in the future. It doesn't matter whether this is a vague, distant desire or a concrete, well-planned assault on a specific marketplace. The end result is the same. You have given the interviewer a good reason **not** to hire you.

Look at it from this point of view: a company will hire you, spend thousands of dollars to train you and teach you their way of doing business. If they are realistic, they don't expect to see a return on their investment for a couple of years, at best. So why should they **give** you all the knowledge you need to set up your own business, probably as a direct competitor in their market?

You would not, I hope, go into an interview and tell the company that you really only wanted to get a few years of experience, so that you could then go to work for their competition at a higher salary. So why tell them essentially the same thing by indicating that you want to set up your own business? Your desire to open your own business should be a personal goal, and one that you would be much better off keeping to yourself.

The fact is that you don't know for sure what you will be doing in the future. If you have a need to brag about this business that will make you an independently wealthy tycoon, find a different place to talk about it. Like in your room, alone, with the door closed.

Turnover of personnel continues to be one of the most costly problems faced by all employers. Don't give that recruiter a reason to believe that you will be just another turnover statistic.

If you are seeking your first career position after serving in the military, you should be aware of what I call the 1 - 2 - 3 - 4 syndrome. Here's what it is:

All during your recent life you have been "promoted" in about yearly increments. You have gone from first grade to sixth grade in elementary school in yearly bounds. Maybe you went to junior high school for three years and entered high school where you were a sophomore, junior and senior. You went on to college where you again followed the one year steps from freshman to sophomore to junior to senior. When you went into the military, you started as an enlisted person or a second lieutenant or ensign and progressed on a well-known timetable to the next rank. (When I was in the military, even these were one-year steps.) The point is that all your life you have been looking at the next step, knowing about when it would happen, usually within the next year. The thought of being a college freshman for three years never entered your mind. It was not part of the normal progression.

You are about to embark on a career progression where, for the first time in your life, you might be doing the same thing for more than the one year you have become accustomed to, and on a very indefinite timetable. In some cases, this might mean you will be filling the same position with the same responsibilities for two or three years or more. If you have not thought about that fact, you should give it serious consideration before interviewing.

The recruiter is looking for reasons why you should not be hired. If the job is one that becomes rather routine, and where the progression is slower than the other positions you might be considering, it is not every recruiter who is delighted to find that you are a person who needs constant change in your environment and responsibilities. You can be disqualified from a good, but slow-progressing position, if you come across as the kind of person who needs to continue the 1 - 2 - 3 - 4 progression on an annual basis.

More importantly, you might think about the effect that a slower moving career might have on **you**. Is this what you want? To what extent will it influence your motivation to do a good job? Promotion will probably not come in one-year increments in any job you take, but there are organizations with clearly structured career paths. Perhaps you'd be more comfortable knowing exactly how long it should take to reach the next rung on the ladder. Seek out those more structured organizations if that is the case.

Above all, understand that the steady and predictable progression you have realized in school and in the military is not going to continue in most career opportunities.

There is a strong consensus among recruiters that hiring decisions should be made only after at least one interview in a non-traditional interviewing environment. In a normal office, desk, chair interview arrangement, people are very tense, have their guard up, and are prepared for question/answer confrontations. Take them out of that environment, and you often see a very different side of the candidate. Most commonly, the "different" setting is at a lunch interview. (Personally, I can learn more about someone in 45 minutes on the racquetball court than in a couple of hours of question/answers.) Interviewers will approach your lunch interview differently, but generally you can assume that the interviewing pressures will be less obvious.

Trying to tell someone about yourself while you have your mouth full of mashed potatoes could be a disaster – or an excellent interview if you are prepared. Since the interview over lunch or dinner is usually scheduled as such, you can expect the unusual circumstances. But it could be a spontaneous event. Maybe the recruiter has fallen behind in the schedule and wants to do two things at once to make up for lost time. Or perhaps you have more positives than expected, and to spend more time with you is the best way to probe them. Perhaps the next person in the interviewing chain is present and a quick second opinion could be accomplished now, over lunch, rather than later, in a different city.

Just the fact that you are not interviewing in the same surroundings as the other applicants will set you into a unique category and you are part-way to achieving that desired "remembered applicant" status you are trying to achieve in all your interviews.

You might wonder what to order. Ask the recruiter for a suggestion or wait until the recruiter orders, and order something similar or lesser in price. Don't order a filet mignon if the recruiter is having a hamburger. Not many business people consume alcoholic beverages at lunch any more, but you might take your lead from the interviewer. Getting zonked on zinfandel won't help you much, but becoming an instantaneous teetotaler because you read somewhere you shouldn't drink a glass of wine at an interview is equally inappropriate, particularly in a social setting. If you don't drink, you don't drink, but please try and keep the self-righteousness out of your voice.

The interviewer is going to pick up the tab, so keep your money in your pocket.

Do not let the situation become a chit-chat. You are there to interview. If the recruiter wants to talk about the weather or mileage flown on the airlines, let it go. But you bring the subject back to interviewing if you see signs of staying away from the subject too long. Remember that you have limited time, and that you are being evaluated in this situation just as if you were back in that office.

Oh, yes. Make sure that you eat your lunch. Don't spend all your time talking.

Group interviews are those situations where there is more than one person facing you in one interview. These group interviews are most often found at the second or third interview stage, although it is not unheard of to have them scheduled at the first stage.

These can be tough interviews, mostly because something happens to the egos of those interviewers when they all get together in one room. They try to out-do each other in showing how tough or probing they can be in their questioning. Also, there is a certain shock factor in being brought into a room where there are several people sitting at a table, ready to have you on a platter (do not confuse this with a lunch interview). There are several pointers here that can be valuable if you find yourself in that group situation.

If you are sitting at a table with five interviewers, there is a strong temptation to listen to and answer the questions on a one-to-one basis. A person asks a question, you answer it looking only at that person. Instead, try directing your answer only partially to the person who asked the question, but also incorporating everyone else at the table, and finally ending up with the original person. This makes everyone feel as if you are including them when you speak, and demonstrates good communications skills.

The questions may come from each person in turn, in an orderly fashion around the table. Possibly the questions will come at random so you cannot anticipate that the next question will come from the next person in line.

Since there is a good chance that this will be your second or third interview with an organization, there is a great possibility that the first one or two interviews were favorable and there is more than casual interest in you. These people will probably be an assortment of your future supervisors, someone who has technical knowledge to quiz you on, or perhaps someone who is presently doing the job you are applying for.

Don't let the group push you around. This does not mean that you should become arrogant and defensive, but it does mean you are there to be interviewed, not chewed out. There is always the opportunity for an interviewer to find "fault" with you no matter how good you are. If a negative, misleading and unhappy interview is ever going to surface, it probably will be in a group interview with several managers and co-workers, each trying to impress the next with their interviewing techniques. Stand up to the questions, but use good judgement.

It is useless for me to tell you to relax and enjoy the situation. The whole group interview scene is a tense one. But do try to relax by telling yourself that you wouldn't have gotten this far if there wasn't some interest on their part. Odds are you have been in front of a qualifying board in the military. This is the same thing, really, and your experience might give you an advantage over a civilian applicant.

Be conscious of the way you sit in the chair, and of your posture. If you have nervous habits, such as wiggling your foot, or playing with your watch or ring, this is the time they will surface. Relax, relax, relax.

Due to the high cost of airfares, hotels, etc., telephone interviews are used more frequently now than in the past. It is to your distinct advantage to be prepared for the unexpected telephone call while you are conducting your job search. Your first impression might be that a telephone interview would be easier than a face-to-face conventional type. For some people, that could be true, but most of you will find that interviewing by telephone is more difficult.

The reason for this is, in part, because you have very few reference indicators when you are talking into a mouthpiece. That is to say, a major part of your normal interview will be influenced by what you read on the interviewer's face, in the body language and by changes in voice inflection or tone. While talking in person, a quizzical expression tells you that you need to give more detail, a frown may mean that you need to defend or explain in a more relevant manner, and a smile means that all is (or seems to be) going well. You do not have these aids when you talk on the phone. At the end of the conversation, you will hang up wondering how well it really went.

Since you should be conscious that the recruiter is also losing **your** facial and body signals in a telephone interview, you will need to make up for it in tonal signals. Your voice should be clear. You should allow as much enthusiasm as possible to enter into it. Slow down. People have a tendency to speak more quickly on the phone than in person. Avoid that if you are able to. You should sound confident and as if it were the most natural thing in the world that you should be telling someone you cannot see why they should hire you.

If you have an audience, clear the room or use another extension. You need to concentrate on what you are doing, just as if you were in the room interviewing with that same recruiter, and you cannot do that with your nephew arm-wrestling with the family dog in the same room.

In a telephone interview, all the interviewing rules apply and all the answers should have full content. You still need to stimulate interest so you will be invited to the next step.

Take note that even a telephone call from a recruiter with whom you have already interviewed, who is simply setting up another appointment with the next person in line, is also an interview of sorts. Your reaction is being assessed. The evaluation of your enthusiasm, response to simple questions, and general attitude is being weighed for the eventual decision the recruiter or interviewer will make about you.

There is one thing you can be positive about when you walk into an interview. When the interviewer eventually asks, "Do you have any questions for me?", the interview is nearing completion. It seems to be a universal method of signaling that the interviewer has asked all the questions that will be asked, and that it is now your turn to clear up any questions **you** have in the remaining time.

Although most sources of interviewing information I have had exposure to in recent years stress the importance of asking questions, it is safe to say that most applicants are so relieved that the interview is about over that they will answer this question with, "No, I don't think so," so they can get out of there. I hope it is equally safe to say that no person who has used this Handbook will **ever** let that happen. We are actually working with two different principles here: the need to ask intelligent questions about the career opportunity, and also the need to ask a question which will tell the interviewer you are interested (see section on "Pursuit of the Employer").

If you have done your research properly, you will have some questions about the organization and the opportunity. Think about it from the recruiter's point of view. There's a limited amount of time to give you information about the company and the job you are looking at. Because of the time restraints, this information must necessarily be somewhat superficial. Isn't it strange that you would be satisfied with that limited input, given the importance of the decision you are both about to make? With only a few exceptions, which we will discuss, there are no limitations on the questions you might ask. I will list a few different questions to show how you can come up with them, but will leave the actual task of forming questions up to you to fit your own situation.

If you have been told that the average promotion time from job A to B is 18 months:

Q: "What is the shortest time it has ever taken?"

If you have been told that you will be paid a small base salary plus a percentage of the company profits:

Q: "What is the company's profit sharing program record?"

If you are told that there are seven plant locations:

Q: "Do you have new plants planned, and if so, how are new personnel selected?"

If you have been told that there is an excellent training program:

Q: "What form does the training take and how long does it last?"

You will note that these questions are designed to make the interviewer be more specific in a positive way. You are showing your interest. This is an important decision for you - you want as much information as you can get. Ask questions your research has already answered to see if there is any difference between what the *Wall Street Journal* has to say and what the company line is. This is **not** the time to turn the interview into a reverse negative interview:

Q: "Why did the Federal Government close two of your plants last year?"

Q: "Why don't you have any women on your Board of Directors?"

Q: "I used one of your products last week and found it inferior to the competition's."

If the answers to those questions are important to you, there will be a time and place to get the information. Along the same line, avoid trite questions or questions which have no bearing on your career pattern in the organization:

Q: "What kind of company car do your sales reps drive?"

Q: "Tell me about your retirement plan and fringe benefits." (This *is* important information, but will be detailed later on in the interviewing process.)

Q: "Will you pay to move my household goods to the new location if I am relocated?"

The time and place to get those kinds of questions answered is in an offer interview, when the company is ready to hire you, and not in an initial interview.

If you blank out and can think of nothing at all to ask, you can get out of that situation by stating, "I have no questions to ask of you now, but I will probably have several over the next few days. May I call you or write to you with them?" Remember, I said you could use that as a result of a blank-out, not as a cop-out. Then go right on to the next phase.

There is one question you **must ask** at the end of each interview you have. It is extremely important and takes a real effort. You must ask:

Q: "What is the next step?"

Not those words exactly, please, but words to that effect. It is not my objective to have hundreds of military parrots around all asking "What's the next step?" What are you trying to accomplish? You are pursuing the employer, certainly. But the next step should also be an important event in your mind. I cannot imagine that you have prepared for this interview, gotten all dressed up, done everything necessary to make it a success, and not have the curiosity to know when and where your next contact with the company will be. To find out that information, you must do something that is a bit uncomfortable for most of us. You must **assume** that the recruiter has interest in you and will want to pursue you. This is where you will show assertiveness in the interview far beyond what you might normally show. Do not leave that room **without** asking something like this:

Q: "I have really enjoyed talking with you, Mr. Jones, and I am anxious to learn more about your organization. When might I be hearing from XYZ Incorporated again?",

otherwise, you may fall into that gray zone you have been trying to avoid. It really matters little how you say it. You could do it by showing interest and concern for being available:

Q: "This has been a most interesting half hour and I am interested in talking with other managers in your company. I want to be home when you call. When might that be?"

If the circumstances of follow-up are obvious, still underline that you understand the procedure, and are interested.

Q: "I know that you are interviewing several people here today and understand that you will make some decisions when you are finished. I am looking forward to hearing the results and the chance of seeing you again."

or

Q: "Yes, I do have one other question. Where do we go from here? When can I expect to be notified of the time and place of my next interview?"

Again, it doesn't matter how you say it. Express honest interest and press for the details. You will get varying degrees of information from this approach. But some information is better than no information and your by-product is the fact that the interviewer knows that you are interested enough in this opportunity to at least ask about it. Final note: this would also be a good time to ask for the recruiter's business card so you spell the name correctly and use the correct title when you send your follow-up note.

If you intend to interview with only one or two prospective employers, you should have no problem remembering the names of the interviewers and when and where you are to meet with them. It is very likely, however, that if you interview with more than two companies simultaneously, you are going to have problems unless you have your act together. It could be quite embarrassing to show up at the wrong place at the wrong time to interview the wrong person for the wrong job. Here are my suggestions.

You need a calendar-type schedule organizer; I'd say one you could fit into your pocket with those important 3" x 5" cards you will always have with you. Into this calendar should go all the information about your scheduled interviews: company name, your contact there, the time of your appointment, and your objective for that particular interview as it will appear on your resume. It would be smart to include the company address and telephone number as well. I doubt that this is an original thought for most of you, but I thought I'd mention it so you would remind yourself to pick up such a pocket calendar the next time you have the opportunity.

The need for a more organized approach will escalate when you are sending out many letters and resumes to different prospects. As we will discuss in the section on "Interview Sources," one letter sent to one classified ad or one letter sent to one company is not playing the averages. You will make many contacts. In order to do the right thing to the right organization at the right time, you will need to set up a suspense file of some sort. Your military experience should have brought you an appreciation for the suspense file, but in case you were not exposed to it there, let's review.

A good example of a suspense file for this use would be a box with dividers for each week in the next six or so months. You should have one card written out for each company you contact, whether that contact is by mail, telephone or in person. The company card moves among the weeks in the box.

If you send a resume and cover letter to IBM Corporation on July 15, the card for IBM is noted with that fact and the name of the person you sent it to and filed in the week you will want to follow-up on that action, say, August 1. File the card in the first week of August. Each week **force** yourself to do all the things you have noted to do on the cards for that week. If you are contacted by IBM on July 25, the name of that company should be fresh enough in your mind that you can pull the card easily.

If your collection starts to grow too large (as I have seen in the teaching profession where several hundreds of contacts might be made), you will want to cross-index for easier reference by listing the companies on a sheet of paper and making a note as to where each card is in the system. This card box should be reserved for those **potential** interviews. IBM Corporation moves into your calendar of interviews once you have established an interview appointment with them. Follow-up with them should be noted on your calendar. If you have a home computer, you could easily do all this with a simple data base program.

As in the case with most of the ideas I will share with you in this Handbook, the system will only work if you make it work. Use it for a few weeks. If you are diligent, you will find it saves time and guesswork. And if, like most people, you find you have 10 or 20 interviews scheduled, it might save an interview, too!

A good part of this Handbook has been devoted to showing you ways to convince the interviewer that you have enthusiasm for the opportunity at hand and how you can show that enthusiasm in your answers to questions and in your actions as you follow up. Now I'm going to tell you to be **unemotional** about the whole thing. Yes, that's right. Be as unemotional as possible about the interviews, the position and the possibility that you might be working for the organization. There are two reasons for the paradoxical approach of being enthusiastic but unemotional.

First, if you get emotionally involved in every advertisement you answer in the classified section and in every resume you send out to a company, you are going to be disappointed so often in such a short amount of time that it will affect your outlook on interviewing in general and your motivation to continue to seek interviews in particular. When answering ads: send it in and record when and where you sent the response in your suspense file and forget it. Resumes and letters should be handled in the same manner. That is what your suspense file is for. **Don't** get anxious and sit by the mailbox every day, waiting for those answers that might never come. Do other things with your time. Send out more resumes or answer more ads.

Second, if you start getting emotional about a particular position that appears to be developing into an offer, your decision is in danger of becoming based on emotion and not on facts or intellect. And unless you make your decision based on what that particular job can do for you, what you can learn and where it can take you, you might be making a poor decision you will regret when the novelty of the attraction wears off.

Easy to say, hard to do. Right! And the truth of the matter is that your final decision will be based on facts **and** emotions. You will have one eye on the hard truths of the job and the other eye on the chemistry that excites you. In that respect, interviewing and accepting an offer are much like getting married. You are emotionally involved but (one hopes) clear-sighted enough to see that things and people change. What you see now is not always what you get six months later. Do yourself a favor and don't get too emotionally involved with your upcoming interviewing opportunities. You can get yourself psyched up just so many times and then it just doesn't work any more. You need to keep that positive attitude, expecting that the company will like you, but also realizing that there is a possibility you may need to interview with several, or dozens, of companies to find the situation that best fits you. Rejection from one position is not total rejection from the entire field. Even the best applicants I have seen have had rejections. They just don't talk about them. Expect it, and when it comes, don't go to pieces.

You should be so lucky!

If you do all the things outlined in this Handbook, you will be prepared for a 30 to 45 minute interview. This is an intense period during which you expect to be evaluated based upon the answers you give to questions and, as we have seen, many other unspoken criteria. Possibly, you will suddenly realize, half-way through the interview, that the recruiter is doing all the talking. The thrust is toward telling you about the history of the organization, the advantages of the company, the excellent reputation, the fringe benefits and popularity of the overall employment situation there.

As you listen to this stream of public relations material, you should also be making a judgement, **because at that point the recruiter has made a decision about your candidacy.** A normal reaction is: "Hey, this person likes me, wants me to come to work for this company, and is trying to tell me how good it is."

You could be dead wrong.

The chances are equally good that the interviewer has decided **not** to pursue you and is using the time to tell you how good the company is for public relations reasons only. After all, there is a given period of time to talk to you, so why not get in a free commercial advertisement? As you have read in the profile on recruiters, a secondary function of this position is to promote good public relations.

The smart approach to the situation would be to assume the recruiter is still undecided about you and to continue to utilize all the pursuit techniques you have learned in this Handbook.

This is a good time to ask career-oriented questions.

This is a good time to ask about the next interview.

This is a good time to be smart and not sit back and assume that you are home free and well on your way to an offer.

At one point, if you are interviewing for a position in business, you will read or hear the words "we promote from within." It is often presented in a meaningful, self-righteous tone and that should tip you to the fact that it is important. But just what does that phrase mean?

On the surface, "promotion from within" means that the company policy is to give a present employee consideration for promotion to a higher position before the organization hires an experienced person from an outside source. If that is as far as it goes, that is fine. However, some organizations large enough to have many promotable employees have such strong internal promotion policies that they could make the statement, "We will never hire anyone unless they start at the bottom rung of the ladder," in effect reserving all management positions for those who have been with the company from step one.

Before we continue, let me stress that most companies would like to promote from within whenever possible. It is a good policy. But most organizations simply do not have the resources to follow that policy for every opening that occurs above the entry level. Many companies have job announcement policies that describe open positions to all internal employees. These companies will generally make sure that no acceptable internal candidate exists before going outside.

A good mix of new talent at the bottom and new talent throughout the organizational ladder is a preferred method for bringing new ideas to most companies. Now, with that out of the way, let's go after the kind of organization that only promotes from within and why that might cause you some problems as a military leader.

Under the surface, that organization with a super-strong internal promotion policy is going to present problems to the applicant with more than a couple of years in the military or in other full-time employment. If the only place you can start is at the bottom, a recruiter must give strong consideration to your salary requirements, your maturity, your experience and your career goals.

Situation: let's assume a recruiter could talk you into working at the entry level in his company. It might not be all that hard to do. You are 32 years of age and you have been in the military for nine years. The experience you have had is flying airplanes, and is not related to the position you are applying for, so you are "entry-level" as far as the recruiter is concerned. Even so, is it really fair to put you in competition with 21-year olds? Just in terms of your effectiveness based on maturity, you should leave most of those kids in the dust. And what about you? Are you ready to cope with the situation yourself? If you are 32, when was the last time you sat down and had a conversation with a group of 21 year-olds, even the intelligent ones you could expect to be competing with? What about the salary? In a company with an iron-clad policy for promotion, you can expect equally structured salary policies. Can you live on that salary after the nine years you have been gaining seniority in the military?

These kinds of questions are going through that recruiter's head and the result is that you could be knocked out of competition unless you have only three or four years in the service and are fairly close to the entry level in salary and age.

If the above applies to you, companies with **firm** "promote from within" policies can be hazardous. Key word: **firm**. Perhaps you should concentrate on making application to companies where there is a good, healthy internal promotion philosophy, but where employment at a slightly advanced position is possible.

While it is not my intent to make this a Handbook for those persons with extensive full-time experience, perhaps you are not one of the few fortunate people who had the opportunity to use *The Interviewing Handbook* when you first left the military and are now reading it after you have had full-time employment and are starting to look for a new position*. This section is for you.

At one point in your interviews, you will be asked why you left or are thinking of leaving your job. Well, you should have a reason, and it should be a good reason, not a shallow one. It doesn't matter whether you were working as an insurance sales rep or teaching school. If you had the opportunity to make a career of it, the recruiter is interested in learning why you decided not to, to be assured that the same conditions do not exist in the new opportunity. There are several things to keep in mind when answering questions about your past employment.

Your previous employer will **probably** be contacted as during a reference check, so the contents of your presentation should not waiver from the basic facts. The word "probably" in that sentence reflects the trend toward not giving out much information about past employees, as a result of legal problems experienced by corporations which have done so with poor judgement. Many companies will only give dates of employment, the job title, and verify the salary if the person requesting the information has the dollar figure. Do not attempt to give yourself a three thousand dollar raise or a paper promotion as you make your presentation.

If you are leaving your position on good terms, your former employer might be an excellent source of recommendation. If

---

* We have had a lot of requests for a Handbook for experienced job seekers; that is a very different interviewing situation. Some day we'll add it to our list of *Interviewing Handbooks*.

you know that the employer will have good things to say about your work habits and personality, tell the recruiter to call directly. Provide the contact information in full. On the other hand, if your exit from the company was one that left some question about your character, don't offer any more information than the personnel office extension unless you are pressed for greater detail.

NEVER, EVER badmouth your last job or employer. This will do nothing for your case in a positive sense, and can do much harm. Since you are giving only one side of the situation, and a prejudiced one at that, any account you give of the terrible conditions of your previous employment will be questioned. It is your word against their's. Unless you can cite a specific example ("my previous boss was convicted of embezzlement and now is in the state pen"), refrain from making unflattering remarks about your past supervisor.

One subject that seems to come up repeatedly when talking with younger applicants with experience in business is politics - as in: "I hated that job. It was entirely too political." Anyone who has been involved with business for much longer than a year knows that corporate business is political.There is much reason for concern if you can't stomach a basic working condition existing in most business situations. Don't use that as an excuse for leaving. Try to avoid giving specific critical reasons for your leaving whenever possible.

Q: "Why did you leave your last job?"

A: "I left my job because there was too much travel involved."

R: "I'm confused. Were you lied to when you took the position? Did they tell you that there wouldn't be any travel when you were hired?"

Suddenly your judgement is under question, not the injustices of your previous situation. It might be better to approach the answer from a different, more general angle, placing emphasis on goals and growth objectives:

A: "I left my last job because I felt that I had outgrown it. I learned a great deal as a sales rep: how to meet people easily, how to travel in an organized manner. But now I am looking for a more challenging position, one which will give me project responsibility in a staff position."

Be prepared to admit a mistake if you have made one. Do not defend your decisions to the point you are being dogmatic.

Q: "If you had the choice to make again, would you still accept the job with Acme Widget as a widget inspector?"

A: "At the time, it seemed like a good choice for me. I needed the income and the job was available. But I can honestly say that I would not make the same decision again, knowing what I know now."

Salary is always a touchy subject. If questioned about it, you should use the same principles discussed elsewhere in this Handbook for the person who is entering the job market for the first time. There is more to a position than money, and you should make that feeling known. Of course, you want to make a jump in pay if feasible for normal progression, but this is not always possible, so don't limit yourself. The most common answer in this situation is:

A: "I was making X dollars at the Acme Widget Company, and I don't want to take a cut."

when it should be:

A: "I was making X dollars at Acme, but I would not be leaving the widget business if I felt there was good opportunity in it. I am open to offers in a wide range of salary, if there is opportunity in the position. I am more interested in making a decision based upon the job than upon salary."

Finally, be sure that your resume includes pertinent information relevant to the position you are interviewing for. Tailor your resume to fit the opportunity by emphasizing on it your production responsibilities for an interview in production, marketing in a position for marketing. Even with a couple of years of full-time experience, you should be able to get it all on a single page.

The legality of testing an applicant seems to be a subject open to a rather wide interpretation. Some organizations will absolutely not allow an applicant to be tested in a sit-down written format. Others with a different view require that an applicant undergo an aptitude test, or a test to help determine suitability for a particular situation.

One area that seems to be in agreement is that technical tests to assure that the required technical skills (typing or electrical engineering as examples) are at a level to perform the job may be administered before an offer is made. The legal key here is that the test is "validated" in a way that shows the results of the test are useful and non-discriminatory.

In my opinion, if you are given a non-technical test, that test should be followed by at least one more interview, regardless of how many interviews you have had with the organization up to the point of the test. If, instead of another interview, you are sent a letter of rejection, you might want to write back asking the reason for rejection. If they were so interested in you that they went to the trouble and expense to do the tests, you can only assume that you didn't do well on the test. Use words to that effect. You might find that the organization will talk to you again and if so, you might be able to change their mind with a good interview. You don't have much to lose by trying. If you have a question as to the legality of rejecting an applicant based on test results, you might call the nearest official State Employment or Equal Opportunity Employment Office.

There is no way you can study for the tests given to prospective employees. They are of the same kind you took in college entrance exams and upon entering the military service. They test a wide range of knowledge including math, vocabulary, spatial concepts, graph analysis, fact retention and so on. The more sophisticated (and expensive) also include an interview with an industrial psychologist. You cannot do much about any of that.

What you **can** do is insist upon good conditions in which to take the tests. If you are taking it in a lobby with people walking around and talking and telephones ringing, politely ask for a private room, or at least a quieter place. It is your right to ask for these conditions if the test is going to be used to consider you for a career position.

At any time after your initial interview, you should be thinking about receiving an offer. As the offer situation approaches, you should watch all developments carefully and start to pursue the offer well before the final interview. If, for example, you asked after the **first** interview:

Q: "I am certainly impressed by what I see here, Mr. Jones. When will I hear from you to set the next interview?"

you should, at a point later in the process when your sense tells you that you are near an offer, be asking this sort of question:

Q: "I am even more impressed now that we have spent these two hours together, Ms. Baker. I think I fit this company. What is the next step toward an offer?"

Not pushy, but with enough emphasis to let the interviewer know that you are interested in an **offer** to go to work for that organization.

Some of you will have extreme difficulty getting up enough nerve to say those words, or words like them. I understand. It is not easy to be this assertive, particularly if you are in the first or second week of your interviewing career, with no offers "in the bag" to rely on. Just let me point out a couple of things.

First, there are some organizations that will not give you an offer unless you ask for it. They are uncommon, but they do exist. If you don't ask for an offer, you don't get one, you just kind of fade away. This is most frequent in sales interviews, where the boldness to ask for an offer is equated with the focus required to "ask for the order."

What about the case where an interviewer is undecided about you after several interviews? Your fate is on the line and you screw up your courage and finally ask:

Q: "When do I get an offer?"

You might hear:

A: "You won't."

At least you have a decision on the case, a decision which is, by the way, probably no different than if you had strung it (and yourself) out over three weeks. But there is also a strong possibility that you will impress the interviewer with your aggressiveness and enthusiasm so that you will receive a more favorable answer.

If you didn't have good characteristics, you wouldn't have gotten this far in the first place, right? This frame of thought includes the same reasoning you used when you told the first recruiter that you were anxious to go further in the interviewing process. It is obvious that you want an offer or you wouldn't be there interviewing, but ask the question. It demonstrates that there is more interest on your part than on the part of the other applicants you are competing with who have never **asked** for an offer.

When your offer does come, it could be a very informal verbal offer, or verbal with a formal written follow-up. It is nice that you have the facts on paper so there is no misunderstanding later regarding your salary, start date, etc. But I would not ask for that offer in writing in this manner:

Employer: "Well, John. We've made the decision, finally, to give you an offer. Here it is . . . etc."

Applicant: "When do I get that in writing?"

Reaction: "Don't you trust me?"

When an employer gives an offer, there has been some significant energy expended. The recruiter or whoever has looked at you, probably spent some money on you, decided you are the caliber of person that fits the situation, has had good indication that you are more than casually interested in the company, and is giving you an offer to **join the team**. This is not the time to bring surprises out of the closet. Nor is it wise to

200

question the integrity of the person giving the offer. You should be flattered. You have made the grade and you have your first real offer. What should you do now????

It is at about this point that many potential problems could surface. Let's examine several of the most menacing, and see what you can do to handle them.

First, let's go back to the point at which you are asking for an offer. The situation was that you had interviewed several different people, and it was about time to either get the offer or move on.

Q: "Ms. Baker, I am impressed. I was impressed with Mr. Jones and Mr. Smith and with the day I spent in the field. You and I have spent a good deal of time together, and I have only one more question. When do I get an offer?"

A: "If I give you an offer, John, will you accept it?"

At this point, John may find himself in a bind, unless he has considered the possibility of this happening. If John has done a good deal of interviewing, or enough to know what the market is like, he may be tempted to answer the question with a very positive strategy.

A: "Once I have the details of the offer, I can answer that question better, but I can tell you now that I have interviewed with several different organizations, and have several different offers, but have not accepted or rejected them until I found out what your company was going to do."

If the situation looks good to you, and the company is one you have put at the top of your list, that kind of answer would be in order. If, however, you are uncertain as to the offers you are going to be getting in the next few weeks, and you do not want to make a commitment, your answer to the question needs to be something like this:

201

A: "Once I have received the offer, I am going to take the details and the knowledge I have gained in interviewing with your organization and sit down with my (wife, husband, father, sister) and make the decision. If I were the impetuous type, I would tell you now that I'd like to accept your offer. But I believe in making decisions of this magnitude (and it is an important decision for me) only after sleeping on the facts, in order to assure that the decision is not emotional, but a lasting one."

That ought to make the interviewer stop pressuring you. I don't think you can call this question a pressure situation. It sounds more like a recruiter's ploy to find out where you rank the opposition. Remember, most recruiters do not like to give out offers which won't be accepted. Someone, somewhere, may be keeping records of those facts, but even more important is the recruiter's self-esteem and sense of pride and purpose. No one likes rejection.

Any high-pressure tactics exerted by an organization for you to accept an offer on the spot or in a one- or two-day period, should be viewed with some suspicion. There is simply no reason you should not be given a week or so to examine the offer and compare it with others.

You might have been thinking of answering that question in this manner:

A: "Once I have the details of the offer, I am going to take the information and compare it to the other offers I am going to get, and I'll let you know whether I'll accept your offer then."

If that thought had crossed your mind, you have forgotten how the recruiter feels about the company. You have been tested and found acceptable, so don't start saying that you are not sure that you want to go to work now, after all this effort and involvement. It is assumed that you have been talking to other companies, but don't tell the person giving you the offer that you want to compare it to others, or that there is any doubt in your mind that this is a top-quality company, the best you've talked to. Leave the impression that you are flattered to get the offer, and are going to discuss it with whomever is going to help you make up your mind, and that you will get back to the recruiter with a decision.

In most cases, there will be no pressure from the organization. You can take as long as you want, within reason. I have heard of situations where the offer was given and the individual was told to give an answer within the next six months. That could be the ideal situation, but not to be expected. More likely, you will be given one to three weeks in which to make your decision. And it is often possible to stretch that time, as long as the organization does not think (or care) that you are stringing them along. All the more reason to be organized and really know what you want, and what interviews you can expect in that period of time.

I said there was usually little pressure from most companies to accept an offer. There are cases where there could be considerable pressure. There are two reasons that come to mind. First, if the recruiter wants to fill the job, he or she knows from experience that the longer a person holds an offer, the less the chance that it will be accepted in the end. Pressure.

Second, there can be internal urgency. Let's say a sales representative quits and leaves the territory open. Every day the territory is open, the company loses money and customers. That sales manager is not going to tolerate extensive delays. There will be pressure to fill the position, so you will be under pressure to accept or reject the offer.

There is nothing wrong with accepting the first offer that is presented to you. Some people would argue the point, but if the circumstances dictate, and you are satisfied that the offer is a good one for you, don't feel that you are breaking an unwritten law by accepting it. It would be ideal to have four or five to choose from, but not everyone has that luxury.

Finally, you should realize that because you must live with the image the employer is forming about you at this crucial point in your interviews, you will not want to drag out the

203

decision unnecessarily. No manager is excited by people who cannot make decisions. If you can make your decision in two days, and you have been given a week, it is a good idea to accept before your time is up. This gives the employer a positive feeling. It shows that you are excited about going to work, and that is a good atmosphere to go to work under.

Accepting an offer is simple. Usually a verbal acceptance is all that is necessary, but it might not be a bad idea to follow up words with a letter of confirmation and thanks, just for the record. In only a small percentage of the cases is a contract signed, and it should, of course, be read and understood completely before it is signed. If you have questions, you should get legal advice. The expense might be a good investment if it prevents misunderstandings in the future. An offer is often made contingent upon both the checking of your references and your ability to pass a physical examination, which is administered at the employer's expense.

A word to the wise. If you are inclined to sample certain controlled substances from time to time, I highly advise absolute and total restraint well prior to and for the duration of your interviewing process. Highly sensitive tests will pick up incredibly small traces of drugs, even marijuana, which will instantly disqualify you and retract the offer. These tests are becoming increasingly popular. Why take the risk?

All too often, an applicant focuses on the goal of receiving an offer, and does not look past that point to actions to be taken when one is finally in hand. Your real objective is to **accept** an offer and you should be prepared to make that decision at the appropriate time. Assess your potential in the job market. Listen to whoever is advising you. Analyze your personal goals. Evaluate the employer's potential. Balance it all out and make **your** decision.

Since this Handbook is generally aimed at the person seeking a first-time career position out of the military, it is unlikely that many will have purchased a house. If you have, there are certain facts to be considered.

You know that if you are interviewing with an organization that is spread out over a large area, it is essential that you be open to any location within that area. If you are approaching the job market with that attitude, your house could become an anchor.

Perhaps you are in the financial position to rent out your property. If so, fine. You have retained that valuable flexibility and could live in your house if you found a position nearby or could rent it out if you were relocated elsewhere. Make certain the recruiter knows that.

More commonly, you will need to sell your house and purchase another at your new job location. Remember that the recruiter is highly interested in the probability of you accepting any offer that is extended. A lot of effort goes into making that offer and a payback is expected. The decision to extend an offer or not could easily hinge on your house "anchor" and that go, no-go decision could be made in the first interview. If there is a strong reason to believe that you will turn an offer down, you won't be allowed to compete in the first place.

The main concern here will be with the **cost** to you of selling your house and buying a new one. How long have you owned it? Long enough to make a profit after paying closing costs of the sale and pre-payment penalties on the loan? What about the loan? What percentage is it? Higher interest rates mean higher monthly payments. Be aware of all costs. Obviously, if you have not thought about this, and don't know how much it will cost you to get out of your present home and into a new one, your "I don't know" won't impress anyone. If you own a house, you would be well advised to find out about these facts **before** you start your interviewing, and not just to impress your interviewers. You may find that the costs involved make relocation financially unfeasible and you need to redirect your career search.

Every time you interview for a position outside normal commuting range of your present home, you will be faced with the same problem, even after you are employed. If you are moving from position to position with your employer, your company will have a specific relocation policy. Some of these policies are liberal and assure that you will not lose any money in the move. But many relocation policies are being restricted by all except the most profitable companies.

This doesn't mean you should live in a trailer rather than a house. But it does mean that you should be aware of the costs involved in relocation once you buy that house and be prepared to discuss them intelligently.

In the next few sections, you will read about sources for interviews. I do not intend the content of these sections to cover all the possibilities, nor am I convinced that anyone can say that there is a particular "best way" to set up interviews with employers. The attractiveness of one method over another is determined by your marketability, who you know, how long you have been looking, what you are looking for, where you are looking, and your personality, to name but a few of the variables.

There is a tendency among applicants to approach the job search on a linear chronological basis. The candidate contacts one employer, schedules an appointment, goes to the appointment, waits for the second interview, receives a letter of rejection, searches around for a second company, finds one, gets an interview, goes to the interview . . . etc. The applicant who follows this pattern is not looking for a career, but for a job, expecting to take the first position offered. A lot of time is wasted **waiting** for something to happen instead of using time wisely, doing everything possible to **make** something happen.

You should interview on a full-time basis. If possible, you should spend forty hours a week scheduling interviews and interviewing. When you are not interviewing, you should attempt to get more interviews. Approach all the sources of employment interviews (with the exception of employment agencies, for reasons we will discuss in that section) at the same time, with a well-organized approach.

It is true that the more people who know that you are seeking employment, the better your chances of finding a career position. Some sources of interviews will never be completely exhausted. But if you want to be thorough in your job search, you should at least touch all the interview sources listed here.

I am critical of college placement offices. If you read the first (1976) edition of *The Interviewing Handbook,* you would see that I had some strong negative feelings at that time about people who ran those placement offices. Their effectiveness was very low back then, and few I had visited (and that was a large number) were worth the trouble.

Since then, many placement offices have improved their effectiveness to a large degree. This change (and it is really a remarkable change - ask any recruiter who has been on college campuses over the past eight or ten years) is driven by the school's reaction to the changing attitude of the student body. Students are interested in careers, more so than ever before. And, they are asking specific questions before plunking down their life's savings, or signing themselves into student loan debt for tuitions, or committing themselves to a specific campus. And one of those questions is: "What is the track record of the placement office? Is my investment going to pay off?"

This new attitude on the part of students caused admissions offices to look to the placement offices. The emphasis in the placement offices changed. Companies were sought out, accommodated and competed for. Simultaneously, curricula were beefed up in many schools so graduates would become sought-after employees, attracting employers to the campus.

Responding to the positive attitude of the campus, employers discovered a breed of student different from those encountered on campus in the early 1980's. These students are serious. Employers are breaking records every year as they sign up to interview on college campuses.

There have always been excellent placement offices. Not surprisingly, most have been at private colleges and universities where the faculty has a vested interest in attracting students – if there are no students, there is no tuition, and, therefore, no faculty. There have been some topnotch public institutions as well, usually because of the energy, savvy, and dedication of the director.

The point of all this is that the attitude about student placement has changed – both on the part of colleges and employers. College placement offices are much more productive than they formerly were.

If you graduated four or six years ago and you go back to your placement office after your military service, you may find that you are allowed to use the resources of the office, but first priority for on campus interviews will go to graduating seniors, in effect barring you from taking part. Since interviews usually take place only in the Fall and Spring, your timing would have to be pretty good anyway, so the office will mainly be a **source** for names, companies, locations of employers, etc.

An exception: many offices have bulletins or current listings for full time-positions, filed by companies looking for recent and not-so-recent graduates. Need I suggest that you look at these closely? Get on the telephone instead of relying upon mail communication for these listings. If a job is open and needs filling to the point a company would go to the trouble of listing it on campuses, you can assume two things: the position does not require lots of specialized experience and it will be filled quickly.

Another thought: as you learn about specific dates that specific recruiters will be on campus, there is nothing wrong with writing or calling that person directly. If you know you have a background that the company has hired in the past, give it a shot. Many recruiters will make room in their schedule to talk with a promising candidate, if not at the campus, then the evening before or after the campus interviews.

In summary, the college placement office is not a super-productive source for the military leader because of the priority given to current students. If you can get on interviewing schedules, you should do well, given your experience and maturity. At the very least, you should be able to get good contact information from your placement office.

No one, except those with strong masochistic tendencies, enjoys rejection. Most people will go to great lengths to avoid it and that's why few of you will utilize this door-to-door approach. It is an invitation for rejection and to spend your time convincing secretaries and clerks you should have a shot at the personnel interviewer.

If you are satisfied with the interviews you have had using other methods, you might limit your efforts using this technique to the few organizations you know about, but which have not surfaced through your other sources. An example would be a company with headquarters in your area, which has not appeared at the campus placement office, advertised in the newspapers, or interviewed at any of your employment agencies. If you cannot get a personal referral through a friend or relative, you must approach the headquarters with your request for an interview. Because so many people fear that ever-present rejection, most prospective applicants will write a letter to the organization rather than visit in person. Writing a letter is a poor second choice if you are able to go to the headquarters or local office personally, armed with your resume and our interviewing techniques. Needless to say, you should be dressed to interview, expecting the best, but unafraid of the worst.

Ask for the recruiter or the personnel department. You will most likely be met by someone ranging from the secretary to the assistant personnel manager. Explain the situation positively and see what happens. At most, you might get an interview, but at least you have the opportunity to leave your resume. If you are asked to fill out an application, use your information cards. If possible, get the name and title of the recruiter or person in charge and write to that person after you have left your paperwork. Express interest and availability to interview at their convenience, and perhaps follow up that letter with another if you have heard nothing in a week or so. Perhaps you could telephone or drop by again, but do not expect too much. If you have the qualifications, and there is an opening, the recruiter will contact you for an interview. The odds of that happening, of your being in the right place at the right time, are not very good, but are successful often enough to go to the effort.

In this case, where you are presenting yourself to someone other than the person who will interview you for the position (e.g., the secretary), the slightest positive reaction is often the difference between getting the interview and getting a "no-thanks" letter.

Picture the recruiter, busy with daily business. From experience it is safe to predict that 90% of the applicants who come through the street door are generally misdirected. You show up, looking sharp and saying the right things to the receptionist. Your resume is good and your application form is exceptionally well done. The receptionist says to the recruiter, "This person just came in, looks good." The recruiter is much more apt to interview you than the person who comes in and "looks bad," right? Keep on your toes. Even approaching the receptionist should be done as if that person were going to go back to the recruiter's office and give a report on you - which happens more often than you might think.

If your interviews have not been plentiful (through other methods), or you are not satisfied with the amount of exposure you are getting through them, use this method to a greater extent.

An additional word of advice on this subject: instead of calling on organizations at random, or only on companies with high public profiles, you should use a more systematic approach. In many areas, an industrial guide is available. A county guide, for example, would be a listing of every business in the county, including the name of the business, the number of employees, the products or a brief summary of the services, and other pertinent information. These may be purchased or used at the library. Just tell the librarian what you need. If you are unable to find a local reference book, call the Chamber of Commerce in your area for information. You may be surprised by how many Fortune 500 companies have operations right up the street.

When you do make contact with someone who will interview you under these circumstances, and are not encouraged in spite of all the techniques you use, don't hesitate to ask the individual if they know about other positions open in other companies for which you might be qualified. In this situation, recruiters are often quite helpful. They keep in touch with their "competitors," see them at agencies, job fairs and other functions and know a lot about the local job market. Even though their current role in your life might make them seem like your enemy, you have to believe they are human beings and, like most humans, have some empathy for the job seeker because they have been there, too.

The major disadvantage of going from door-to-door is the fact that you do not know which opportunities are available behind the doors you are knocking upon. For this reason, the objective on your resume should be a little more broad than we discussed in the section on resumes. This is only allowed when you are applying in person! Your objective might read: "Entry-level position in marketing, sales, or marketing research, leading to managerial responsibilities." This is broad enough to encompass more than the basic sales job, but specific enough to discourage the company from considering you as an applicant for their janitorial staff.

Ordinarily, such a wide-ranging objective would be unacceptable. But when you are presenting your resume in person, you will count on your impact and personal interviewing skills to overcome the interviewer's objections, if any, to this seeming lack of focus.

This method is not always fun, but it can be extremely productive. If you can access the right people, they will be impressed that you are applying directly. Not surprisingly, the more productive it is, the more fun it becomes. Keep your spirits up by realizing you are doing something most other people will not and, therefore, have the chance of beating them out for that offer.

When done correctly, this method also produces some interview invitations. The first thing you should keep in mind is that major organizations receive hundreds of resumes each week and during the times of year that college graduates gear up (September to November and February to May), the number is often in the thousands. If you are a person of average qualifications, I'd advise skipping this method all together. There are just so many resumes floating around that an employer responds only to those entry-level resumes which reflect outstanding applicants. Now I realize that few of you consider yourself average, so let's look at how you might approach this method.

Your first step is to write a resume. Since you could be considering as many as a couple of hundred mailouts, you should have the resume printed from your camera-ready copy. It will be relatively inexpensive and will produce a good, clean copy. In contrast to the personal, door-to-door approach discussed in the last section, you must have a fairly **specific** objective on your resume or it will have little chance of being taken seriously. You are trying to sell yourself to someone who has never seen you, asking for an interview. If you don't fit into currently open positions, your resume is generally dumped.

Your next step is to construct a universal cover letter for your resume, explaining that you will interview for any position for which you are qualified in your area of interest. This keeps you open for positions related to your objective, but does not present a hazy catch-all. This letter might also explain when and where you will be separating from the military and other pertinent information not included on your resume, including any "free" relocation information you are aware of.

The cover letter should be typed but, if you will be sending out hundreds, you could have them printed and type the names and addresses of the employers at the heading. Please do yourself a favor and use the **same** typewriter or typeface for the printed letter and the addresses you add later. To use different typefaces is really tacky. Even at best, this method does not look very impressive so you might check out the cost of having your cover letters created automatically with word processing program tied to a laserwriter. The resume and cover letter should be stapled together to increase the odds they will remain intact in the process of being opened and delivered to the desk of the individual concerned.

Before you address the envelopes, you will need a list of places to send them. There are several methods of developing a list. You can use the same sources as discussed in the section on "Knocking on Doors" for establishing leads, or mail to likely organizations listed in the publications available at your college placement office or library, or search old want ads for leads (see next section). You could easily run one to two hundred copies. This may or may not be worth the time and expense, depending upon the number of interviews you have already scheduled. Again, if an organization receives these mailouts by the truckload, they are going to take the time to contact only the most qualified candidates for specific and immediate needs. Your letter must be a good one, and the material on the resume must be and look exceptional.

Expect many totally ignored requests for interviews, lots of letters of rejection, and a tiny fraction of positive results. You keep track of the mailouts (as we discussed earlier) with 3" x 5" cards and a suspense file. If you are of the inclination, follow your letter with another after a reasonable period of time (three weeks) has elapsed with no response. Keep the tone positive.

If you are exceptionally well qualified, or just plain lucky, the mass mailout is a worthwhile way to get interviews. On the other hand, if you are exceptionally well qualified, you will have more than enough interviews from other means, and can restrict your mailout to only those excellent organizations which are inaccessible to you through other means.

Mass mailout is a popular waste of time among job seekers. Mailing out those resumes and writing those letters does keep a person busy, so many applicants hide behind this busy work instead of going out and knocking on doors or using other methods for getting interviews. Then, if they cannot get a good offer, they protest that they have mailed out three hundred letters and gotten little response and therefore, there are no jobs available. It gives them something to bitch about and justification for the fact that they are still unemployed. But **you** know that mailing out letters blindly is at best a semi-productive means, and a passive means at that, for getting interviews for entry-level positions.

I don't want to confuse the issue, but I feel I must add that this mass mailout method is productive in a specific case. Let me use an example to illustrate.

You are stationed in Seattle and want to move to and work in Chicago. (Why you'd want to do that is beyond me, but anything is possible.) You have sufficient cash to finance an interviewing trip and Aunt Florence, on your mother's side, has said that you can stay with her and Uncle Fred in Mt. Prospect and even use their '72 Pinto for around-town transportation. You are set.

Go to the library, and find out which companies are headquartered or have substantial operations in Chicago and do a mini-mass mailout. Now here is the key: very few companies are willing to spend the money to fly an entry-level sales applicant with a B.A. in Art History and four years of pushing troops in the Marines from Seattle to Chicago for an interview, sight unseen. You could probably find several such applicants on the streets of Chicago. Why go to the trouble to import them? However, if you write your cover letter explaining (in the first few lines, so it gets read) that you are coming to Chicago **at your own expense** to interview, few recruiters would be so insensitive to their public relations role that they'd turn you down if there was a possibility of a match. Urge a timely response to your request. After all, you have plans to make for this trip. Call those companies you really want to see if you have not heard from them. By being willing to put your money where your resume is, you may impress a recruiter to the point of getting an interview.

In summary, if you feel you should use mass mailouts to get interviews, use it. Do not fool yourself, however, into believing that you are really doing the complete job search sitting at home, stuffing envelopes and sticking on stamps.

Classified advertisements have a bad reputation. The story goes that letters sent in response to ads are dropped into a bottomless pit, seldom to be heard from again. An answer might come back in the mail, but it is always a job interview to sell life insurance, regardless of the wording in the original ad. The reputation is possibly deserved in some cases, but I have serious doubts anyone can overlook the classifieds as a source of interviews, particularly when such a small investment of time and money can pay off such large dividends.

What is involved here is a system of reading ads on a regular basis, keeping tabs on when you wrote to whom, and a good follow-up system. The ability to read an ad to find out what it really seeks is also important. Setting up the organization system is relatively simple. If you wanted, you could cut out each ad and tape it to a 3" x 5" card along with the pertinent facts.

The key to success in the system of ad response is consistency. If you pick up the paper every once in a while to look in the classifieds, you are counting on luck to do most of the work for you. The ads must be read **every day**. Make it a part of your routine. If you subscribe to the local paper, or the paper for the area in which you wish to work, this is really not difficult at all, and might take all of twenty minutes a day. If you need to go out and buy the paper daily, you are probably not going to read the ads every day, so consider taking out a subscription. It is a small investment. If you are stationed at a military installation in the middle of nowhere, you might consider this means very seriously, writing to the ads of interest, explaining your situation complete with estimated time of arrival upon separation from the military.

Why must you read the ads every day? Because it is expensive to advertise in a newspaper! You would stagger at the price of a medium-sized advertisement in a major newspaper. Often employers advertise only on the weekend, which is when the majority of people look at ads. If the employee being sought is hard to find, the greatest exposure possible is desirable. Conversely, if an employer has a position that is really wide in appeal and requires only general qualifications, the advertisement may appear only in the middle of the week to cut down on the response. If you think it is a treat to screen five hundred resumes, you won't appreciate why a smaller response is desirable.

If you are not reading the ads every day, you might miss the ad that really applies to you. You will quickly become familiar with the recurring ones, and will ignore them to concentrate on the new ones which might hold some interest for you. To save time, you could prepare ten or more packets, ready to be mailed to the ads as you discover them. A packet would consist of your resume and a cover letter, typed or reproduced by a good quality copier or by a printer. The two pieces of paper are put in an envelope, and the packets are ready to go to the advertised addresses. This is quite a time saver, for it eliminates the need to find an envelope and write a cover letter every time you want to respond. All the contact information is recorded on your card for filing.

In the event there is a telephone number to call, you should have all the information you might believe will be asked of you at your fingertips. Your resume will contain most of this information, but you should be prepared for what will possibly be a telephone interview. Needless to say, you will have your calendar notebook handy to schedule an appointment.

Many advertisements are "blind." That is to say, they ask for response to a post office box, or to a box in care of the newspaper. This method is used to discourage applicants from calling the company (and bugging the recruiter), or to search for an applicant without the employees of the company knowing that there is an opening (as in the case where someone is about to be fired), or to simply avoid having to respond to all those resumes. Do not be discouraged by this method. You are going to invest a few cents in a postage stamp and the cost of the newspaper, but it may prove to be worth it. The important thing is to establish a system of response and to remain consistent.

Most applicants will send in a response and forget about it unless there is an immediate reply on the part of the advertiser. That might be what you want to do for your own mental health. However, if you see a particularly appealing ad, one you feel you are qualified for, but receive no reaction from your initial letter, you might invest another few cents and the time it takes to write a letter. An example:

"On June 10, I answered your advertisement in *The Daily Blatt*. I know you must have received many responses because the position you described was particularly interesting, at least from the information you offered. I feel I might be well qualified for this position and would enjoy the opportunity to discuss my qualifications with you in more detail than a letter and resume allow. I am available at the following, etc..."

Use a positive approach in your pursuit of this employer. You will set yourself above the average.

This same thought could be applied when you receive a letter of rejection, which usually says something like this:

"Thank you for your interest. Since receiving your application, we have interviewed many qualified applicants and have selected a candidate who matches our needs (may or may not be true). We will retain your application and contact you if a suitable position develops." Signed, Mr. Jones' secretary.

99.9% of the applicants will give up at this point. If you sent Mr. Jones a note, re-expressing your interest in his organization, and in spite of the lack of a suitable opening at present, told him you hope he will keep you in consideration for future or related opportunities, are you wasting your time? Possibly. But if I have learned anything in the employment market, it is this: persistence pays off. Nice persistence, not aggravating persistence such as telephoning three times a day. If and when Mr. Jones does have an opening fitting your qualifications, who do you think he is most likely to remember? You, the note writer, or the applicant who has just sent in one letter or resume in response to the ad? In my experience, the prize goes to the person who asks for it, not to the person who sits back and hopes that it will fall into his lap.

When people tell you there are no good positions advertised in the classifieds, agree with them and continue to watch the ads daily. Many employers who use other sources for attracting applicants, such as employment agencies and college placement offices, also advertise in the newspapers. Your interview can be productive whether you apply through the classifieds or through any of the other sources. Ads cost big bucks and employers are getting a return on the investment, or else the classified pages would be blank.

You must read the ad itself carefully. Many are worded quite ambiguously. Be aware that the advertiser might be inserting "qualifications" into the ad to discourage response from thousands of applicants, but might not hold to the ten years experience demanded in the ad copy. It might also be a case of the recruiter not really knowing what is out there and shooting too high in the requirements. If you come close, apply. If you are in the right place at the right time, you could get lucky and be interviewed. If you are invited for an interview and use the techniques you have read about in this Handbook, you will have a better chance of getting an offer than the next person, even if that person has all the required experience.

On another note: out-of-date newspapers can be a valuable source of names, addresses, telephone numbers and jobs. Your library probably has old editions of your Sunday *Blatt* on microfilm. Go though the past year or so and glean the **contact information** published there. Send resumes to them if they look promising, but this is a case where you should ignore the blind ads mentioned earlier. There is a limit on the amount of time the "box" stays open, but if you go back two weeks or so, you just might find a good one. Send your packet. The recruiter might not have found anyone the first time around and welcome the chance to talk to you.

Classified ads are a method which should be incorporated into your total job search. For the small amount of time and money involved, it would be a good place to start. Don't be discouraged by some of the lofty requirements you see in the ads. Most established companies are well aware of the positive attributes of a military leader. Maybe you'll get lucky.

The first time I wrote this chapter, it was called "RELATIVES, FRIENDS, AND THE TAXI DRIVER." The concept has caught on (although I can't take all the credit for that) and is now called "networking" - same chapter, different title.

When I interview an applicant who has told me that a father or uncle or sister is employed in a management position with an organization, I question whether this relative has been asked about employment opportunities. Too often I get this response: "No. I want to do this myself, on my own."

My reaction is not favorable. I can relate to the emotion, but not to the logic that goes into such a statement. If you are faced with a job market which is extremely competitive, and do not use every avenue available, you are letting pride or ego stand in the way of one of the best sources of employment - the personal contact. I am convinced this is the method used in the majority of job placements.

I am eventually going to say that the more people who know you are looking for a job, the greater your chances of having a choice among several offers. But (before I say that), let's look at an example where you have a father who is involved in business and has worked for a company for several years.

Many parents are wise enough to avoid influencing their children in the choice of career direction or specific job opportunity. Your father might be holding back his assistance until **you** bring the subject to him. Does anyone have more opportunity to evaluate your work performance than your parent? Believe me, he will not refer you to someone he feels will not approve of your qualifications. On the other hand, he might know some people, either in his own organization or in companies he has contact with, who need an outstanding military person for entry-level positions. If there are common military ties there, all the better. Use his influence to get that initial interview.

Rest assured that the minute you start that interview, you will be judged on your own merit. There are few employers who are going to **hire** you as a favor to an acquaintance or employee if you don't fit the qualifications. On the other hand, there are few employers who will not give you an **interview** as a favor to an acquaintance. Some companies even have a strong program for encouraging such referrals for interviews. And that is all you need - an interview.

Approach your parent (or whomever) on that basis. Make it understood that you are not looking for an easy track to an offer, and expect to compete with other applicants for any available positions. Then use the techniques you have learned here to interview as if it were any other employer, taking nothing for granted. Even if your relatives are not in the particular business you wish to seek as a career, there is a good chance they may know other people in the same field you are trying to enter. Use these people. Do not let the opportunity go by just because you are unwilling to ask the question due to pride. How proud will you be if you are still unemployed six months after you start interviewing?

The situation is about the same with friends you might have who are employed. It is likely you will have several friends who are already established, even if you are just getting started in the process yourself. Someone who has worked for an organization for a year or six months probably knows **much** more about the avenues of approach to that and other organizations than you do. There is even a chance that your friends will receive an incentive from the company for initiating interest on your part. Also, these friends may have a list of names from their interviews when job-seeking and would let you have the list. Get out your address book and look up your friends who left the military before you did. Remember how you said, "We've got to stay in touch," and "Look me up whenever you're in Atlanta"? Well, now's your chance. If they are employed, pump them for information about the who, what, when, where and how they got employed. Do it. They may appreciate the opportunity to help!

Older friends of the family will possibly help you even more, particularly if they are well-established in their field.

Again, unless they are particularly close to you and your family, they will probably not offer such advice unless asked directly. It's human nature.

Everyone you and your family knows should be aware you are looking for a position, and their help should be solicited. Never pass up the opportunity to ask an informed person if they know someone they can refer you to, if only for an "informational" interview to learn more about a different company. Since you are doing all you can in the other methods of scheduling interviews, your request for assistance will not be interpreted as looking for the easy way out. In fact, you should never pass up the opportunity to tell someone you are on the market.

What this translates into is establishing a "network" of sources, which are keeping your interests actively in mind. The wider you spread the word, the greater the chance that fact will reach the right ears. Getting a good offer is not, as you may realize by now, simply luck. More accurately, it is lots of hard work and just a little bit of luck. Most often you have to make that luck work for you.

A good employment agency is one of the best sources of productive interviews. A strong statement, but a very true one, particularly if you underline the word "good."

As you may have heard, there are agencies and then there are agencies. The best ones attract the top companies as clients and have a scope that can be extremely valuable to you in terms of time and effort saved in your job search. You should not walk into any agency with your eyes half-open, so let's take a close look at what an agency can offer, as well as some of the limitations of the agency approach.

Basically, we can break down all agencies into two kinds - the ones where you, the applicant, pay the fee (which are rare, at least in major metropolitan area); and the ones where the employer pays the fee. Since this fee can be a substantial sum, you will no doubt make this one of first things you find out about an agency. In my experience, there is no case where a military leader with (probably) a college education needs to pay a fee to an agency for job referrals.

The two classifications of agencies are further specialized into employment specialties. There are agencies which specialize in secretarial and related office positions, accountants and finance people, computer and electronic data processing positions, engineers, sales representatives, experienced sales reps, managerial personnel and many other classifications. Also, some agencies specialize in particular kinds of applicants - college graduates, military officers, experienced personnel, temporary help, etc. With all these variables, you can see that selecting an agency to assist you in your job search is not as simple as picking up the telephone book and dialing a number. In fact, a major part of the bad reputation that seems to surround the agency business in general, is reputation generated by applicants who have not made the effort to select the appropriate agency for their needs. They are naturally disappointed when the agency cannot produce interviews for them or tries to steer them in a different direction.

Instead of just choosing an agency from the Yellow Pages, you will want to do some initial screening and legwork to make the best selection. Most of this preliminary effort can and should be done over the telephone, eliminating those agencies which don't fit your needs. You can save a great deal of time by calling to determine the basic requirements of the agency, their specialization, their geographical scope, who pays the fee, and how to go about registering for their services. In short, you will be selecting an agency where you have the most in common with their general placement and recruiting efforts. To help you understand how an employment agency works and make the most out of the opportunities you may find there, let's take a closer look inside an agency.

Most agents (or "counselors") work on some sort of a contingency commission basis. When they place an applicant, they receive a percentage of the fee charged to the company for that placement. Although they refer to themselves by many different titles, you should realize that **successful** agents (the ones you really want to be associated with) are excellent salespeople. They are successful for one reason - they do not let things happen, they **make** them happen. Agents in an employment agency are selling under some unusual circumstances when compared to other sales personnel. This is a sales situation where both the buyer (employer) and the product (applicant) are variables which can change from moment to moment - quite a different situation from selling steel beams or dill pickles. People selling those products know that they can rely on their products to remain relatively constant, and concentrate on the one variable, the buyer. It takes a skillful and sensitive employment agent to bring the two variables together to an accepted offer, and a good agent is paid well for those efforts.

Some agencies are franchise situations with wide variations in quality and results from office to office. If your friend in Boise had good results with one branch of a particular agency, a similarly successful experience at the Duluth branch is not guaranteed.

The fees that are negotiated between the employer and the agent are generally based upon your starting salary, but can have a wide range, up to 35% or more of your annual salary. It might occur to you that agents would want to place you in the highest paying jobs possible, since their fees are related to your starting salary and the higher your salary, the higher their fee. Not necessarily so. Even though an agency is paid per applicant, they are evaluated by their client, the employers, on several other factors which are not really related to you accepting the most highly paid job opportunity. The ratio of follow-ups per interview (how many applicants must be interviewed before a qualified one is found), the ratio of offers extended to those follow-ups, the acceptance rate of offers, the progress of those new employees in the company, and how long the employee remains with the company are critical factors. Any agent who pressures an applicant to accept a particular job simply because of the higher fee, will eventually lose credibility with client employers and not be in business long after that.

Which brings us to an interesting point. If you get all settled in with an agent at the agency of your choice and discover that he or she has only been working for two weeks, you may need to reconsider your choice. Ask what kind of experience he or she had before coming to work there. Ask about the client companies. If it sounds as if you have better contact resources and know more about the selection process than the agent, walk. You need an agent who is an asset, not a liability. Find one you are comfortable with, whose company has made an investment in training, and development.

A successful agent works in an environment which provides access to many different job opportunities and can probably match you to a number of them. The number of matches, from the agent's viewpoint, is best kept small. If you are allowed to interview all of the companies in the files and consequently receive offers from three of them, that means the agent must disappoint two clients, since you can accept only one of the offers. Again, credibility with clients is at stake, and if they are disappointed often enough, they will go elsewhere with their business.

Let's look again at the fact that an agent's success is based on **making** things happen. Payment comes from filling client needs with qualified applicants. If you approach an agency and request assistance with your job search, giving them specific information on what you want in a position, and the agent does in fact move quickly into action, securing job interviews for you which result in job offers, you are suddenly in a decision-making posture. You owe yourself and that agent a decision on whether or not you want to accept that opportunity. Different circumstances will create this need for a decision: the company has an opening which it needs filled now, not in 30 days, there are other applicants who will be considered for the position if you do not take it, or a training program will start shortly. Friend, you asked for action, and now you've got it! You'd better be prepared to make up your mind! For this reason, I strongly suggest that you know what you are looking for before you approach an agency for assistance. If you do not have that direction, that knowledge of what you are looking for, look into other sources of interviews before approaching an agency. If you want to experiment, look around, or just practice interviewing, do so elsewhere. Realize that a good (there's that word again) employment agency may move quickly, and you will need to make important decisions in relatively short periods of time.

All this brings up the question of why employers use employment agencies for entry-level positions in the first place. One of the main reasons is that companies sometimes cannot meet their staffing requirements with other means. Their college recruiting efforts have not filled all the openings (often companies give out two to three times as many offers as they have openings to compensate for this factor - sort of like an airline overbooking a flight), or possibly they cannot wait until the college applicants graduate for the openings to be filled. An agency will be contacted if the company has not been able to recruit applicants through in-house recruiting procedures, or has no other sources of good applicants. The amount of time and money involved in running an advertisement in the classified section of the newspaper, waiting for the response to the ad, and then interviewing the applicants who respond (many of whom may not be what the job requires), is very high and an investment some companies hesitate to make, particularly if hiring is an infrequent function of the person doing the interviewing. A good agency will supply that same company with several applicants who are in the same ballpark as the requirements and in the long run it could be less expensive to hire someone through an agency. Notice the time factor running throughout these reasons. If the position is one which generates dollars for the company, such as a sales position, every day it is open represents lost income to the company - a great motivator for the company to find the right person and do it fast. As an agency applicant, you will not have the luxury of three weeks to make your decision.

There is one other reason employers use agencies, and it has to do with the section on recruiters as public relations people. If you remember, I painted a really terrible picture of what could happen if a recruiter alienated a large percentage of the persons interviewed. Suddenly the company is losing sales and going under, just because of the applicants who were rejected. That was an exaggerated example to demonstrate a point. The fact remains, especially within the consumer industries, that each rejected applicant might very well hesitate before picking that company product off the shelf. If a company wants to avoid this negative feeling, they would be better off interviewing a small number of qualified applicants and rejecting a smaller percentage of them than interviewing a large number of applicants from an advertisement and rejecting a larger percentage of those who apply. An agency can fill this public relations need as well, and could well be worth the price to the employer if the recruiter normally does high volume interviewing.

I am including this section in your Handbook because several acquaintances whom I asked to read the manuscript made the same comment: "You talk about sales and sales leading to management often. Why the emphasis? I don't think most of the people who read the book are going to want to be in sales." I feel the subject is worthy of some discussion.

A quick review of last Sunday's classified ads revealed that approximately 40% of the jobs offered were under the heading of "sales." That classification could be broken into two different categories: "sales" and "sales leading to management." The difference is that "sales" is generally a non-developmental position and "sales leading to management" is a management training ground. Remember, there are very few absolutes in anything involving employment, so it cannot be said that either of these classifications is incapable of becoming the other. The "sales leading to management" position could have promise, but little real promotional potential and be, essentially, a "sales" position.

Should you consider a straight "sales" position? Since you are probably a college graduate, oriented toward the management you got a taste of in the military, you probably would be disappointed with a sales job if it involved only **maintaining** a particular segment of business with little opportunity for advancement. I'm not knocking it. Sales is where the big money is made and a sales rep could be rewarded extremely well, if successful. Example: a real estate salesperson probably has little interest in managing a group of sales people, particularly if the money, and lots of it, is in sales. If you look at the staff of a real estate office, it is likely that you will find the most motivated and most highly paid people out in the sales force rather than inside managing the office. The office manager has different skills. In addition, you have all heard of successful sales people who have made enormous amounts of money, even fortunes, and have never set foot in college. Often, these people are money-motivated rather than management-motivated.

You are probably directed more towards management if you have gone to the trouble of getting a degree and additional specialized training in leadership. The question to you is not: "What is your objective?" (because we assume it is management), but "How are you going to achieve that goal?" As you might anticipate, there are few organizations which will take you right out of the military and make you a manager in their structure. You must prove your abilities to manage yourself and the work you are assigned before reaching the point where you will be allowed to manage other people. For many applicants, "sales leading to management" is an excellent way to show that ability. Let's look at the actual job.

All your life you have been exposed to sales representatives. They came to your home at night selling encyclopedias, they sold you used cars, they try to sell you magazine subscriptions. It is natural that you should have a stereotyped image of this profession. The foot-in-the-door approach is not what most of us would willingly pursue as a lifetime career. Throw that stereotype out of your mind for a minute, if you will, because there is a whole stratum of sales professionals you have never come into contact with in your life. And the job they do is quite different from the job of the stereotyped, door-to-door sales rep.

I would like to introduce you to Jane, the sales representative for a well-known, long-established paper manufacturer, and the link between the company and its customers. She is well-trained in the methods and details of the company's service and product. Before Jane was allowed to talk to her first customer, she was given training which enabled her to anticipate problems that might be encountered in the field. She might have spent several weeks in a classroom-type environment, then gone into the field with an experienced sales rep who showed her how it was really done and who introduced her to the ropes. At one point, she will be given responsibility for a territory, which might be in the neighborhood of a quarter, a half, a million, or ten million dollars worth of the company's business. She has an objective to shoot for, a good amount of training under her belt, and a manager who is there to help if she gets into deep trouble. She is almost always on a base salary with some sort of incentive to keep enthusiasm and creativity high. Her primary objective is to increase business, but she is also judged on her ability to solve the problems of customers, help work out their advertising, find out why there is a delivery problem, check quality, and thousands of other little details. As the representative and advisor to customers, Jane is essentially doing a responsible job managing a sales business. The fact that she is a woman does not surprise anyone nowadays, even in a formerly male-dominated industry.

It is possible that she might use some of the techniques of the door-to-door solicitor. There is a go-out-and-find-them attitude in any position where a person is responsible for expanding business. By a large margin, most sales reps are involved in commercial, institutional or industrial sales rather than a direct sale to the consumer. This professional sales person sells a product to a customer who changes that product in some way, and then resells it to the ultimate consumer. A good example would be our friend Jane who works for a major paper company. She does not go door-to-door with a ream of paper under her arm asking people to buy a truckload of paper. She is more likely involved in selling paper to printers or to distributors, and those businesses then turn around and resell the paper as printed books or stationery once they have added value to it. She will have to compete with other representatives of other paper companies for customers, and it is in that competition where the aggressive approach to new printers and stationery stores becomes necessary. The majority of the calls she makes will be on established customers, and it is unlikely she will be expected to expand business until she is well in control of those established customers.

There is a lot of excitement and personal freedom involved in being a sales rep, and the lifestyle attracts an enthusiastic and gregarious personality. Few companies set the hours of their sales representatives rigidly so there is often a degree of flexibility which you cannot find in other jobs. As long as you are doing your job, meeting the needs of your customers and expanding your business at an expected rate, you will be relatively unsupervised.

Remember, if you are interested in sales leading to management, you would only be doing this job for a limited period of time until you have proved your basic abilities. Since you can prove yourself in many different kinds of work such as (production management, advertising, accounting, or underwriting), why is sales a better way to prove your ability?

Look at it this way. If you are an office supervisor and have a number of individuals working for you, at one point you may need to recommend someone for a promotion. Let's say you narrowed it down to two people, both of whom have identical skills. They can do a business analysis at about the same rate, they both know all the administrative procedures, and even their time with the company is about the same, so you cannot use that as a qualifying criterion. Which of these two people are you going to recommend for promotion?

If you think about it, or other instances where you have been in similar situations, you will realize that there could be many **subjective** considerations in making that decision. Who has the best attitude? Who dresses more neatly? Who is on time more often? In most non-sales or staff positions, the criteria for promotion are often heavily weighted in these subjective areas.

In sales, there is also a degree of subjectivity in deciding who should be promoted. Any decision involving people will have some level of subjectivity. At the same time, the sales manager has the opportunity to use some very **objective** data when deciding who is doing the best job and deserving of a promotion. That manager need only look at a monthly computer printout. The sales rep who is at the top of the list in terms of improvement of territory over the same period last year is doing the best job and should be considered for promotion if this performance has been consistent. If the other skills are in place, that higher performer should be given the chance to show others how to do it as their manager.

This is a very definitive method of measuring the ability and progress of each of the sales representatives. Reconsider the administrative situation. How do you measure the ability and progress here? By the number of pencil erasers the individuals use? It is often difficult to make an objective evaluation of the staff member's ability. If the thought of competition under the spotlight of that computer printout appeals to you, take a closer look at sales leading to management.

I am **not** suggesting that you become a sales representative. I **am** suggesting that, since sales positions make up about 40% of the present job market, you should take a closer look before rejecting it based on the stereotype you have built over most of your life. I am convinced there is an enormous need for well-informed, conscientious sales representatives. There are so many inept ones out there that a smart person with a service attitude, who really does what was agreed to, and delivers on time, is practically assured of success in sales.

The next logical step for the sales representative is sales management, but there are many other positions open to the person with sales experience in an organization. Sales experience provides exposure to many sides of the business not readily available to the person who goes to work in an office staff job. Direct exposure to the company philosophy, customer attitudes, the problems in the field, the competitors' products and/or services, and the marketing strategy are important experiences for all managers at all levels.

As we mentioned earlier, recruiters are often taken from the sales force because of this knowledge. The public relations department, advertising, marketing, distribution, customer service and other fields are also possible career paths for the experienced sales representative to follow. In fact, if you look at the top managers of most marketing-oriented organizations, you will find a period of sales experience in almost all of their backgrounds.

The attractiveness of the sales position becomes more apparent if you realize that a degree in marketing is not always a prerequisite. The person with a non-business degree can often find a good position in sales. The evaluation of sales applicants is usually based more upon personal qualities and demonstration of abilities than upon formal schooling. As a result, interviews for sales positions with good companies are often the toughest interviews to take. If you are an engineer, you can show your qualifications on paper, and be considered an immediate candidate. You cannot get a degree in sales.

Sales interviewers use criteria which are not identifiable on paper, and need to be explored in person, using every interviewing technique in the book.

If you feel you come closer to fitting the requirements of a sales representative for a company with a good, competitive product, I hope you take a look at the opportunities available in the field. There are many jobs available in sales, partially because of the stereotype you probably had before you read this section. It discourages most applicants and they are unwilling or unable to accept the fact that there are many sales situations which are extremely professional and totally outside of that door-to-door image. In fact, I feel strongly that a person does not go into sales because of an incapability do anything else, but rather because of a capability to do **everything** else.

Recently, in Dallas, I had a conversation with a young man, about 25 years old, who would remind you exactly of a frustrated non-commissioned officer from your military experience. "Paul," he said, "There are a lot of "gods" out there. Managers who say 'do it my way'. They think that they are always right. It's 'my way or the highway' with them, and they are driving me nuts."

Embodied in that statement is the essence of what is changing in the work world today, and it constitutes the greatest challenge you face in your transition to a civilian career - especially into business, but I think eventually into any field of work.

The forward edge of change and evolution in the work world is impacting the manner in which you (and all managers) will be **required** to interact with your "subordinates." "Required" is a strong word, and it will cause many of you to scoff at what you will read in this chapter. That doesn't surprise me. This material has been scoffed at by the best - it's expected reactive human behavior. There is change in the air, friend, and the forces, economic and otherwise, which are in play are unstoppable. They will change our life and work environments drastically and that is difficult for anyone to accept.

What's going on? Background:

Not too long ago (I can remember it distinctly) in the 1960's, the words "Made in Japan" carried the connotation of cheap, flimsy, copied. A laughable attempt by a beaten power to compete with the major industrial manufacturers. Fifteen short years later, in 1975, the Japanese were kicking our ass in Detroit. That trend has continued, not just in cars, but on other fields of economic battle: computers, TV's, cameras, electronics, microwave ovens, motorcycles, stereo equipment, pianos, steel. There are people in the corrugated box industry who speak in reverent tones about a Mitsubishi corrugator - major industrial capital equipment. Others talk about Kubota lift trucks and tractors, Makita power tools, and, for God's sake, Ashai beer. What happened during those fifteen years?

Good question. Lots of people have asked it. Important people. People with the power to change what is going on. They discovered a lot of things, among them that these products were just plain outstanding. They worked. Well. And if they didn't for some reason, the (Japanese) company backed them up. The company seemed to care that people got their money's worth when they made a purchase.

Another funny thing was discovered. You can **triple** your advertising budget for an automobile, put beautiful blondes in convertibles, substitute the car for sex, apple pie and the American flag, but if it doesn't run well, people won't buy it. Instead, they hear from their friends about an inexpensive car that gets you from point A to point B reliably, and costs less than the Detroit vehicle. At one point in our history that car was called a VW, in the 70's and 80's it's been called a Honda, or a Nissan. And they still have a major hand going into the 90's Similar stories abound in other products and industries.

What made these Japanese products so good and such an attraction? No one will deny that originally the cheaper labor costs came to the bottom line as a lower price, which has an inherent attraction. However, at the height of the Honda feeding frenzy of the '70's, people were buying Hondas which had been loaded out by (American) car dealers with magnesium wheels, stereo tape decks, pin striping, sun roofs, sun shades and tinted windows. Prices were higher than many U.S. cars, but people bought, because there seemed to be **value** in a car (or stereo, or computer, or TV, etc.) that worked and lasted.

Yes, prices were attractive, but beyond this lower price issue, the quality/value issue loomed like a grizzly bear– or more appropriately, a setting sun. "How do you get quality and value into a product?" was the question asked by those important people. The answer started the change that will affect you so significantly and is deceptively simple: you involve the people who make the product in a way which gives them the true sense that they have a stake in that product, and which connects that stake to the fate of the company and their own job and job security. In short, you give them responsibility.

Now's the time. Cue the scoff.

Let's review the traditional method of management. Guy or gal comes to work. Boss tells him or her what to do, guy or gal does it, eats lunch, does it some more, goes home, wakes up the next morning, goes to work and does more of the same. If he or she has a difference of opinion, it's usually, "I'm the boss, I know best. Do it my way or hit the highway." That system exists. Would you do that? Maybe. Particularly if you needed the job to pay the mortgage. But would you like it? Not much. Would you flourish and develop under that system? Give a damn about the customer or product or management? Doubtful. Fulfill your highest potential? Never.

Enter "participatory management", two energy-packed words which cause grown men to scoff.

"Now wait a minute. Do you mean to tell me that those clods out there on the factory floor (or my subordinates in the office, secretaries, clerks and bottle washers) are going to *participate* in the management of what is going on in my business?" Yep.

Let me ask you something. Do you remember the last time you did something that contributed to your team's effort? The time you personally did something that really advanced what it was you and your fellow military leaders were trying to accomplish? Do you remember the sense of accomplishment? How it tasted (you could almost chew it) and how powerful it made you feel? Sure you do. Do you think you have a monopoly on that sense of accomplishment?

Everyone with a working brain can experience that feeling as well as develop the thirst to do it over and over and over again. Recognition and sense of belonging rank right up there with sex drive and sense of security. The Japanese found a way (some say with the help of an ignored Harvard Business School study) to tap the individual's creative energy through his or her involvement. Result: products of higher quality, better value, and excellence.

This "discovery" created a flurry of books on the subject. You'd go broke buying every book with "excellence" in its title. Some companies started experimenting with quality circles, quality of work life and other systems. Many experiments failed. A few survived, and some of these with fantastic results. More companies tried and succeeded, but still many failed, some in instances which were torpedoed by managers and supervisors who saw this participation thing as a threat to their security and personal way of life. Other companies were successful and kept the source of their success quiet, like a trade secret.

Today, as we enter the 1990's, it is obvious to those in touch that this revolution is not a passing fancy. A couple of reasons stand out. First, companies or plants which have done this successfully have achieved such a competitive advantage that interest in the concept has been generated in the same way a nuclear reaction takes place. Second, and I think this is the real power behind the movement, once you give people a taste of the sense of worth, contribution and unity that comes with participating, they'll never settle for the old traditional way again. In the not so distant future, companies will either be participatory or be out of business. If you need further convincing, and most will, do some library research. (I'll save you some trouble) Start with "A Tale of Two Worlds" starting on page 101 of the June 16, 1986 edition of *Forbes*.

The force is moving. Yes, I know this is a simplistic account, there are other factors contributing to Japanese success: the yen, trade agreements, different cultures, different markets. But that force out there will not go away. Furthermore, it is no longer "Japanese." In its universal essence it says that a supervisor or manager does not have all the answers, does not always know the right way and the "right way" could be done better with a little input from the folks who actually do the work. And in that is your greatest challenge.

You must convince industry that you, a military leader from the epitome of a structured work environment, are capable of not just functioning, but thriving in this environment of de-emphasized rank and status.

Maybe that won't be necessary. Maybe you'll go to work for a traditional "I'm the boss" company which wants autocratic leaders and this won't be an issue, yet. But if you are interviewing with companies which are currently functioning under this new approach, you'll need to convince employers that you **could** function in that kind of environment. If you are a scoffer, it may be tough. If you have had your interest pricked, possibly you are open-minded and non-traditional enough to accept that there may be a better way.

At the beginning of this Handbook, you did an exercise in which you answered nine questions in outline form. No doubt you have thought about those answers from time to time while you progressed through the Handbook. Let's go back and look at those answers closely.

1. "Where did you grow up?" You might have found this to be a superficial question when you were first asked it, but I hope now you have a different perspective. It is probably one of the first questions in any interview, and you should have answered it with depth (section 20), instead of just giving an answer the interviewer could have read on your resume or application form.

2. "What are your favorite hobbies?" Again, you could have been tempted to give a surface answer. But now you will not only answer that question with depth, you will answer it with empathy (section 22) and relate some of the characteristics of the particular job to your hobbies, using those that are most applicable.

3. "What attracted you to this particular opportunity?" A difficult question to answer if you did not have clear objectives, but your self-analysis (section 2) helped you define your objectives and gave you concrete examples to use in presenting your case. If you are able to, you could show that this particular opportunity is an extension of your normal progression and direction (section 24).

4. "What are your salary requirements?" I'd be willing to bet you gave a definite number. You won't now that you have been exposed to the material on the three finites (section 30). Nor will you answer question 5 "What geographic location do you want to work in?" with a definite location, unless you are applying to a local or regional organization.

6. "What do you consider to be your most outstanding personal attribute?" You won't have any trouble coming up with that quality, now that you have done your homework, and the one

you do come up with will be presented with depth and empathy so that the employer can relate to it.

7. Telling the interviewer about any sales experience you may have had should no longer be equated to selling encyclopedias and brushes door-to-door. In case you had not noticed, you have been selling yourself to someone most of your life, and presentation of your resume (section 7), application form, and personal contact with your future employers is in itself a sales situation. You might have opened up your eyes a bit and seen that not all sales jobs are door-to-door (section 54), and the word "sales" might not carry that negative connotation for you any longer.

8. If your past commander could change you in any way, no major changes would occur because you recognize this as a misleading question (section 25) and an effort to get you to give damaging information in the interview (section 23).

9. What you have gained most from your military experience is, above all, related to the job you are interviewing for (section 24), whether that was a gain in maturity, skill or leadership.

Neat, huh?

This list of questions is drawn from experience in actual interviews. Many of these questions are asked in different ways, in all kinds of interview situations. The purpose of including them in your Handbook is not for you to sit down and memorize answers to them. As you certainly know by now, that approach would be counter-productive. I do feel you should say the answers aloud with the words you might use in an actual interview. I am not trying to program you, but only familiarize you with the basic concepts you will want to project in your presentation.

This can be an extremely worthwhile exercise, particularly if you have someone else ask you the questions as you answer onto a tape recorder for later analysis. If you have a good friend, or a family member such as your spouse, who is willing to help you, have that person read the questions as you answer them on tape. I don't care how well you know that other person, you will try to make yourself look as good as you can and the situation will closely resemble the pressures you will experience in an interview.

When you are finished, listen to the answers and look for places you can improve your presentation. You might find that outlining the major points of your answers will help you organize your thoughts and remember the main ideas you want to get across.

I have included several questions which are not job related and would be considered illegal in most situations. You may be asked these questions anyway, probably more out of ignorance than malice, so I include them for your information. The difference between legal and illegal is fairly well defined - questions must be **job related** to be acceptable. Sometimes the context is important.

1.   How does your spouse feel about your possible career with our organization? (illegal question)

2.   How did you make the decision to go to college?

3. If you had your last ten years to live over again, what would you do differently?

4. Of all the courses available to you in college, why did you choose the major you did?

5. What do you like to do when you have free time?

6. What was your favorite subject in college?

7. What job would you choose in our organization if you had your choice?

8. If I were to approach your previous employer, what would they say is the single thing you need to improve most?

9. What are your five strongest personal qualities?

10. Why do you think you'd like to work for this organization?

11. At what point in your life will you consider yourself successful?

12. Tell me about yourself.

13. Is there any location in the United States where you would not be willing to work?

14. If you start at this entry-level position, how far would you expect to have progressed in five years?

15. Do you like to travel?

16. Why did you (didn't you) complete a Master's degree?

17. Who do you think is going to win the next (major) election, and why?(borderline legal)

18. How did you finance your college education?

19. Is there any good reason you feel I should hire a person who has pushed troops (or whatever) for the last four years to be a . . . (fill in your objective)?

20. Why should we hire you?

21. Of all the variables from position to position and company to company, which is the most important to you?

22. What are your salary requirements?

23. If you can, tell me the one person, historical or contemporary, you would hold up as an example of what you'd like to be.

24. What does your father do for a living and how has that influenced your career decisions? (illegal question)

25. You are the worst applicant I have seen in seven months. Who told you to come here in the first place?

26. Why don't (or do) you want to work for us in a sales capacity?

27. Give me a good example of when you have led a group of individuals to accomplish a goal.

28. Give an example of when you influenced someone to do something you wanted them to do, but which they did not want to do.

29. What is the greatest accomplishment you have achieved in your lifetime?

30. What are your shortcomings?

31. What is your closest friend's occupation? (illegal question)

32. If you had all the money you needed, what would you do with your life?

33. Is there any experience in your past we can relate directly to your present objective?

34. What are your concerns about managing minorities? (or as a minority) What are your concerns about managing non-minorities, and what problems do you anticipate (depending on context, could be illegal)?

35. What has been your greatest disappointment in life?

36. Use one word to describe yourself and tell me why you chose it.

37. What extra-curricular activity did you enjoy most in college?

38. Have your brothers and sisters also gone to college? (illegal question)

39. Define the word "aggressive."

40. What is the worst thing I could find out about your past?

41. If you were hiring people at your level, what would you look for in their past records that would indicate a successful future?

42. Is there anything we have not discussed about you and your qualifications that you'd like to bring up now?

You are now just about as prepared as a written aid can make you. However, I would feel remiss if I did not leave you with a couple of thoughts I believe are important.

A few sections back, I compared the search for a career position to the act of searching for and buying a house and said there were many parallels. I'd like to use that analogy to make one more point.

Everyone who has thought of buying a house has a distinct mental picture of what that house looks like. It's a dream house on the side of a hill, maybe an acre or so, with a white picket fence and apple trees in the yard. Most people come to realize that they must change the image of their "ideal" house early on in their house hunting as they face the reality of mortgages, taxes and interest rates. But some people refuse to change; they continue to look and look, and all the while houses appreciate in value while the money they have in the bank sits there waiting to be used, probably gathering interest at less than the inflation rate.

What I'm saying is: don't do the same thing with a job search. Almost everyone must compromise at some point during the interviewing process, making changes in that "dream job." If you search endlessly, you are probably going to find that you have come no closer to your ideal after an extreme amount of time than you were at the end of a reasonable amount of time. Just about every position is what **you** make of it and few jobs are what they seem to be in the interviewing process. If you delay, you may lose valuable time that could have been spent building and growing in your job.

Finally, I am going to tell you something that might sound unusual to some of you, but I feel it is based on good judgement.

Feel free to tell your friends about this book. Indeed, I hope you will. Referrals are our best advertisements. But be careful how you approach an employment situation with the information that you have read *The Interviewing Handbook.*

Why? Let's put it this way: if you sat down to play poker with several people who were out to take your money using a marked deck of cards, would you tell them you could read the marks just as well as they could? The interview situation is really no different than that poker game. You are interviewing with recruiters who are out to make a decision about you, a decision which could be of great importance to your future. They are going to use certain methods and techniques not readily known by most people, and now **you** know a lot about the methods!

Much of what you have read here is common sense, concentrated into one place. But I want you to notice those misleading questions, those negative questions, and all those marked cards while you interview. Note how you might have answered those questions without this knowledge. Ask yourself whether that agency or those company recruiters really need to know that you have learned the rules of the game in *The Interviewing Handbook*, before you blurt out that information.

If you are successful in applying the principles you learned here, you will stand head and shoulders above the next applicant with similar (and even better) credentials. Let the recruiters believe those extra inches are yours and yours alone.

Good luck in your interviews!!

We are always interested to hear about our readers' interviewing experiences. If you have comments which you feel might assist us in upgrading future editions, please write.

the
Interviewing Handbooks